CONFLICT REALISM
UNDERSTANDING THE CAUSAL LOGIC
OF WAR AND WARFARE

*This book is dedicated to my wife, Heather, my daughters
Elizabeth and Caroline, and my sons Abrams (Abe) and Phoenix.
I appreciate all your support, encouragement, and love.
Thank you for helping make this book happen.
I love you all.*

Conflict Realism

Understanding the Causal Logic of War and Warfare

AMOS FOX

Copyright © 2024 Amos Fox

First published in 2024 by
Howgate Publishing Limited
Station House
50 North Street
Havant
Hampshire
PO9 1QU
Email: info@howgatepublishing.com
Web: www.howgatepublishing.com

All rights reserved.

No part of this publication may be reproduced, stored in a retrieval system, or transmitted in any form or by any means including photocopying, electronic, mechanical, recording or otherwise, without the prior permission of the rights holders, application for which must be made to the publisher.

British Library Cataloguing-in-Publication Data
A catalogue record for this book is available from the British Library

ISBN 978-1-912440-53-5 (pbk)
ISBN 978-1-912440-52-8 (hbk)
ISBN 978-1-912440-61-0 (ebk - EPUB)

Amos Fox has asserted his right under the Copyright, Designs and Patents Act, 1988, to be identified as the author of this work.

The views expressed in this book are those of the author and do not necessarily reflect official policy or position.

Contents

Figures, Tables and Boxes vi
Foreword vii
Acknowledgements ix
Abbreviations xiii

Introduction 1

1 A Taxonomy of Western Military Thought 14
2 Conflict Realism 29
3 The Paradoxes of Modern (and Future) Armed Conflict 43
4 The Principles and Inverse Principles of War and Warfare 69
5 On Urban Warfare 98
6 Sieges and the Consequences of the Urbanization of Warfare 125
7 On Attrition 158
8 On Precision Strike Strategy 172

Conclusion 195
Bibliography 202
Index 213

Figures, Tables and Boxes

Figures

Figure 3.1	Tracts and Phasing over Time	67
Figure 4.1	Minimum Goals, Ukraine	74
Figure 4.2	Entropy Cycle	77
Figure 6.1	Victories by Type of Action	132
Figure 6.2	Victories by Type of Actor	134
Figure 6.3	Victories across Time, by Action	136
Figure 6.4	Victories by Actor, over Time	136
Figure 7.2	Causal Mechanism for a Form of Warfare	168

Tables

Table 1.1	Framework for Warfare Matrix	11
Table 3.1	Logic 3: Command to Force Comparison	51
Table 4.1	Principles and Inverse Principles of War	72
Table 4.2	Principles and Inverse Principles of War	82
Table 6.1	Victors in Post–Cold War Sieges	126

Boxes

Box 4.1	How to Apply the Principles of War	80
Box 4.2	Framework for Thinking about the Inverse Principles of War	81
Box 4.3	Framework for Thinking about the Principles and Inverse Principles of Warfare	95
Box 4.4	A Holistic Look at the Inverse Principles of War	96

Foreword

The challenge of identifying the likely shape of wars to come is an endeavor that challenges every generation. It is an undertaking fraught with uncertainty. This ambiguity is driven by the many variables that affect the study of future conflict which includes location, belligerents, technologies, intentions, ideologies, and weather among many other factors. Compounding this extraordinarily difficult task is that nations, and their military institutions, must then undertake the difficult task of ensuring that the right force structures, technologies, people, and ideas are invested in and that they are deployed the right places at the right readiness levels to deter conflict – or win them.

Imagining the shape of future conflict was given a boost in the nineteenth century through two different but somewhat related developments. The first was birth of military fiction in the 1870s. Starting in Britain and proliferating to the continent and across the Atlantic to the United States, military and civilian authors wrote hundreds of stories that sought to better understand how the new technologies emerging from the second Industrial Revolution might impact, and indeed define, the wars of the future.

A second and more profound development was the birth of modern military theoretical studies. The best known is Carl von Clausewitz and his work, published posthumously, which was called *On War*. The core ideas of this book, particularly about the relationship between policy and war, the remarkable trinity and the notion of the fog of war remain highly influential in both western and eastern ideas about war and warfare. Others followed Clausewitz, among them Jomini, Bernhardi, and Mahan.

These first modern military theorists laid the foundations for a new generation of military theorists in the twentieth century. Again, seeking to define how new technologies might impact future war, while also seeking strategies to avoid the statement of the western front during the First World War, these theorists developed new and evolved ideas about war on the land, at sea, in the air and in the amphibious liminal area between.

But, as Amos Fox proposes in this new book, there has in recent decades been a decline in the influence of military theorists. The root cause of this in his view is the interplay of institutional concepts, doctrine, strategy, and plans, which he describes as the tetrarch of western military thinking. The impact of this tetrarch has been to build an almost impervious barrier to new ideas and concepts about the phenomenon of war, and the conduct of human warfare.

Conflict Realism: Understanding the Causal Logic of War and Warfare provides a fresh review of the state of western military theory. Applying both military history and a profound understanding of modern conflict, Fox exposes the weaknesses in current thinking and planning for future war. Further, the book proposes that the challenges of fighting in urban areas, problems with siege and precision warfare, and the likelihood of different forms of proxy warfare must drive a reinvigoration of military thought.

To achieve this, military institutions must incentivise the emergence, development, and influence of a new generation of military theorists. These rare but very valuable people might be found both in and out of uniform.

With recommendations for policymakers, military practitioners, and scholars, this is a book that should be read widely, pondered at length and debated at length. I found the pages within this book intellectually exciting and challenging. They provide an invigorating rationale for nothing less than the wholesale reconsideration of many of the ideas that underpin the twenty-first century profession of arms.

Major General (retired) Mick Ryan
Senior Fellow for Military Studies, Lowy Institute

Acknowledgements

I have a handful of people who I would like to formally acknowledge for their role in helping make this book possible. My goal with the writing and presenting on military thought that I have done throughout the past decade has been to breathe the life back into the field of military theory. Military theory – as I argue in this book – is a field of study that resides between International Relations and formal institutional military concepts and doctrine. Military theory has rested nearly dormant for the past 75 years or so as the field of International Relations moved into its own and states began to more rigidly formalize concept and doctrine development. There are a few exceptions, such as crisis points along the way when neither International Relations scholarship, nor state military institutions had appropriate ideas readily available to address emerging military problems. The 2003 invasion of Iraq is one such instance. When the war did not go according to plan and a ruthless insurgency developed, it caught the United States military and its partners such as the British Army flatfooted. A few theorists, such as John Nagl and David Kilcullen, existed in the margins of military thought. Yet, the punctuation of crisis – the United States military's inability to adequately address the insurgency – allowed the United States Army to quickly grasp onto counterinsurgency theory and help offset its cognitive deficiencies and get its forces on the ground operating in a more appropriate way. This should be a lesson for the defense and security studies communities, to include state militaries. Military thought and military theory is too important to be left to International Relations scholarship, state militaries, and think tanks. This book is my attempt to help rectify that problem and provide the community with a book of military theory to help address the challenges of twenty-first century armed conflict.

I would like to thank Peter Roberts for getting this book off the ground. Peter has been one of the few, and earliest, scholars to help address the absence of military theory in the twenty-first century. While serving as the Director of Military Studies as the Royal United Services Institute, or RUSI (my favorite think tank), Peter led a team of tremendous political-military analysts such as Jack Watling, Nick Reynolds, and Sidharth Kaushal in breathing life back into military thought. Perhaps of greatest importance during this time, Peter introduced RUSI's *Western Way of War* podcast which addressed military thought not from an International Relations or military doctrine lens, but from a theoretical approach. This gave theorists such as myself a significant platform from which to begin debating military theory in an open forum, as opposed to in the back channels of military journals and little-read academic journals. Thanks to Peter, I appeared as a guest on the *Western Way of War* podcast early in its run. Subsequently, Peter had me on his other podcasts to help debate important ideas in military thought. In the end, Peter helped coordinate this book with Howgate Publishing, illustrating Peter's important role in breathing life back into the growth of military theory in the twenty-first century. On a personal level, Peter's help has been instrumental to helping me by providing me avenues to expressing my own thoughts on the subject. Few people along the way have been as helpful to me as Peter, and for that I am eternally grateful. Thank you, Peter.

Next, I want to recognize Frank Hoffman. Frank has been another theorist whose work emerges at just the right time to help provide states and militaries with the bridge between cognitive gaps in defense and security problems. At the same time, Frank has been instrumental for my growth as a military theorist by providing thoughtful criticism, encouragement, and networking to help me improve my work along the way.

Next, I want to thank Vladimir Rauta. First, Vlad was a proxy war scholar whom Frank Hoffman put me in contact with as I first began iterating my thoughts on proxy war scholarship and its related military theory. Vlad provided valuable assistance to me along the way, helping me better understand not only proxy war scholars and scholarship, but also in improving my own analytical skills. However, Vlad later served as my Ph.D. advisor. In that capacity, Vlad continued to help me improve my own analytical skills while guiding me through the Ph.D. process while a student at the University of Reading. Vlad's assistance to me as a scholar has been invaluable and his imprint can be found throughout this book.

James (Jimmy) Kimbrough, IV also deserves recognition. Jimmy served as my Squadron Executive Officer when I was a Captain assigned to the 11th Armored Cavalry Regiment (11th ACR). During my first six months 11th ACR, I worked for Jimmy as his assistant. Later, I commanded D Company, 1st Squadron, 11th ACR. Throughout that time, Jimmy offered many insights into military theory that I did not know existed. While serving as Jimmy's assistant, he helped weekly professional development sessions with the officers in his staff section. He taught us many lessons on military thinking by introducing us to military theory, but also theoretical ideas from other fields of study, to include the social sciences. These professional development sessions – truly meaningful events in which Jimmy operated as a peer and not a superior officer – were pivotal learning experiences for me. Later, while commanding D Company, I had less interaction with Jimmy as it pertained to organized professional development. Nonetheless, we would occasionally talk 'shop'. One such occasion was following the National Training Center's (NTC) first decisive action rotation in March 2012. This rotation was the NTC and United States Army's first force-on-force rotation following nearly a decade of counterinsurgency-oriented rotations. During that rotation, D Company had tremendous success fighting against the 3rd Infantry Division's 3rd Brigade Combat Team. Following the rotation, I was talking with Jimmy and expressing many of my thoughts, feelings, and sensings about fighting 3rd Brigade during that rotation. At one point during the discussion, Jimmy told me that many of the ideas I was expressing sounded like ideas he had read in Robert Leonhard's book *Fighting by Minutes: Time and the Art of War*. He recommended that I pick up a copy. Leonhard's *Fighting by Minutes* is perhaps one of the best military theory books of the twentieth century. Having taken Jimmy up on that recommendation, I purchased *Fighting by Minutes* and it fundamentally changed my appreciation for military theory forever. That conversation with Jimmy set me on the path that has lead to the publication of this book. Great mentors like Jimmy are a diamond in the rough and institutions should find ways to curate the careers of Jimmy Kimbrough. So, Jimmy, a huge thanks from me to you – you probably do not fully appreciate your impact on me as a young officer, but it was tremendous. Thank you for being a great mentor.

Furthermore, I want to thank Major General (Retired) Patrick Donahoe. I have known Major General Donahoe since he was Lieutenant Colonel Donahoe and he was commanding 1st Battalion, 67th Armored Regiment, 4th Infantry Division at Fort Hood (Cavazos), Texas. Major General Donahoe has been one of the few leaders in the United States Army to advocate for, and openly curate, non-traditional, non-institutional thinking in relation to military thought. On more than one occasion during my Army career, Major General Donahoe offered me jobs on the spot, without caveat. Major General Donahoe also had me as a presenter at the Maneuver Warfighting Conference at Fort Benning (Moore), Georgia when he commanded the Maneuver Center of Excellence. In his post-Army career, Major General Donahoe has contributed to helping support junior Army officers by taking important positions at Columbus State University and Columbus Community College in Columbus, Georgia. Patrick Donahoe has been a true stalwart for junior Army officer education and the enhancement of military thought. For those reasons, as well as always being a supporter for me and my work, I truly appreciate Major General Patrick Donahoe.

This book has been a labor of love and it would not have been possible without the people noted within this acknowledgement. I hope you enjoy this book and that it makes the reader think. To be sure, the reader does not have to agree with my thoughts, findings, and recommendations. However, I hope that this book contributes to twenty-first century military thought and theory in a meaningful way by forcing dialogue. Moreover, as a International Relations scholar and military theorist, I hope that this book helps further accentuate the gap between International Relations scholarship and institutional military concepts and doctrine. Thank you.

Abbreviations

ALB	AirLand Battle
Ar	Aggressor's resources
At	Aggressor's available time
Ao	Aggressor's operations
CJFLCC	Combined Joint Force Land Component Command
COFMs	Correlation of Forces and Means
COG	Center of Gravity
COIN	Counterinsurgency
D	Dominance
DOTMLPF-P	Doctrine, Organization, Training, Materiel, Education and Leadership, Personnel, Facilities, and Policy
Dr	Duration of enemy contact
Dr	Defender's resources
Do	Defender's operations
Dt	Defender's available time
Ea	Enemy action
Ft	Frontage (i.e., how much space a combatant must cover on a battlefield)
IAC	International armed conflict
IHL	International Humanitarian law
ISF	Iraqi security force(s)
Ir	International community's response
JDAM	Joint Direction Attack Munition
JADC2	Joint All-Domain Command and Control
LOAC	Law of Armed Conflict
MDO	Multidomain Operations
MRB	Motorized Rifle Brigade
MRR	Motorized Rifle Regiment
NIAC	Non-international armed conflict
NGO	Non-government organization
US	United States
UK	United Kingdom
OIF	Operation Iraqi Freedom

OIR	Operation Inherent Resolve
PGM	Precision Guided Munition
Pc	Points of enemy contact along a front
Qf	Quantity of a combatant's force
Re	A combatant's on-hand resources
Rp	Resource replenishment
Rx	Resource expenditure in an adversarial environment
S&T	Science and Technology
SU	Self-sustainment
Ti	Time
UAS	Unmanned aerial system
V	Victory

Introduction

Attempting to predict the future of war is a hot topic in the defense and securities and international relations fields today. Prior to the Russo-Ukrainian War's flair up in February 2022, many observers focused their attention on the supposed lessons of 2020's Second Nagorno-Karabakh War. In that conflict, Azerbaijan and Armenia fought in a tightly compartmentalized theater dominated by mountainous terrain, making movement on the ground slow, telegraphic, and restricted. The terrain allowed Azerbaijan to maximize the impact of its new armed and surveillance unmanned aerial system (UAS) fleet against the land-based Armenia military forces, resulting in a lopsided Azeri military victory. In the fervor to welcome in a new revolution in warfare, many analysts referred to this conflict as the first war won by drones.[1] Nonetheless, both belligerents tended to violate one of the fundamental truths in warfare, which is to always fight with combined arms. As it were, Azerbaijan predominately fought from the air with their fleet of UAS, while Armenia fought with old Soviet tanks and a feeble air defense system that could not effectively counter the Azeri's UAS threat.

Taking a moment to recover from the emotional exhilaration of potentially glimpsing the future of warfare, a balanced perspective should surface. Comparing the general conditions in which the conflict was fought – small theater, mountainous terrain, canalized road network, urban operating environments – it is easy to understand how Azerbaijan quickly defeated Armenia. While neither side systematically employed combined arms, Azerbaijan's ability to operate beyond the reach of Armenia, while

---

[1] David Hambling, "The 'Magic Bullet' Drones Behind Azerbaijan's Victory Over Armenia," *Forbes*, 10 November 2020, available at: https://www.forbes.com/sites/davidhambling/2020/11/10/the-magic-bullet-drones-behind--azerbaijans-victory-over-armenia/?sh=8a9d0e5e571b; John Antal, "The First War Primarily with Unmanned Systems: Ten Lessons from the Second Nagorno-Karabakh War," 2021, available at: https://higherlogicdownload.s3.amazonaws.com/AUVSI/7bc57aaa-ae26-4c7a-93f9-512dc4a1bca0/UploadedImages/Ten_Lessons_from_the_2d_Nagorno-Karabakh_War_by_John_Antal_2021-03-08F.pdf.

Armenian land forces struggled to move through the region's mountain road networks allowed Azerbaijan to put on a dazzling display of warfare that appeared to galvanize support for drone warfare, and yet again signal the death of the tank, and the anachronistic character of human-centric land warfare.[2] Many policymakers, scholars, and military practitioners alike view the Second Nagorno-Karabakh War as a punctuation in armed conflict, a revolution in military affairs, and a 'mic-drop' event, if you will. More level-headed onlookers, on the other hand – perhaps those that better versed in military theory, and the study of tactics and operations – see little more than the application of bad tactics and the deterministic impact of terrain on military operations in situations like Nagorno-Karabakh.

The Second Nagorno-Karabakh War, plus years of proxy wars and counterinsurgency in Afghanistan and the greater Middle East, and almost no peer-competitor interstate armed conflicts resulted in many ostensible policy and strategy mavens to suggest that conventional, mechanized warfare was dead.[3] Further, this crowd tended to support the hypothesis that the Second Nagorno-Karabakh War was a 'mic-drop' event and the future of warfare would be a drone- and autonomous system-centric affair.[4] As Antoine Bosquet suggests in his provocative piece, *The Battlefield is Dead*, war is entering a post-mechanized epoch in which networked drones will silently patrol the skies above the field of battle, seeking terrorists, command posts, armor, or supply lines to systematically strike and destroy any target with precision guided munitions.[5]

According to Bosquet, and other likeminded individuals, warfare in the future will be fast, networked, and robotic. Many of these conceptual

---

2 Benjamin Bremlow, "A Brief, Bloody War in a Corner of Asia is a Warning About Why the Tank's Days of Dominance May Be Over," *Business Insider*, 24 November 2020, available at: https://www.businessinsider.com/drones-in-armenia-azerbaijan-war-raises-doubt-about-tanks-future-2020-11; Alex Gatopoulos, "The Nagorno-Karabakh Conflict is Ushering in a New Age of Warfare," *Al Jazeera*, 11 October 2020, available at: https://www.aljazeera.com/features/2020/10/11/nagorno-karabakh-conflict-ushering-in-new-age-of-warfare.
3 Jahara Matisek and Ian Bertram, "The Death of American Conventional War: It's the Political Willpower, Stupid," *The Strategy Bridge*, 5 November 2017, https://thestrategybridge.org/the-bridge/2017/11/5/the-death-of-american-conventional-warfare-its-the-political-willpower-stupid; Sean McFate, *The New Rules of War: Victory in the Age of Durable Disorder* (New York: William Marrow, 2019): 25-42; Antoine Bosquet, "The Battlefield is Dead," *Aeon*, 9 October 2017, available at: https://aeon.co/essays/how-the-bloody-field-of-battle-made-way-for-precision-drones.
4 "The Azerbaijan-Armenia Conflict Hints at the Future of War," *The Economist*, 8 October 2020, Available at: https://www.economist.com/europe/2020/10/08/the-azerbaijan-armenia-conflict-hints-at-the-future-of-war.
5 See John Antal's *Seven Seconds to Die: A Military Analysis of the Second Nagorno-Karabakh War and the Future of Warfighting* (Havertown, PA: Casemate Publishers, 2022).

disciples cheerfully refer to themselves as Futurists. Further, these Futurists use the hottest taxonomical cliches, embracing what Alexander Montgomery and Amy Nelson refer to as possibilistic thinking; that is, thinking about the future of war through the lens of possibilities, not probabilities.[6]

As I have already noted in my other writings on the four schools of thought in modern military thinking, the problem with this type of analysis – and much of the Futurist school of thought – is that it is assigns linear causality between a small number of data points and often neglects to view a conflict in context, or with reservation for the importance of reality.[7] Somewhat echoing this idea, scholar Patrick Porter writes that:

> Futurologists assumed intense, overt, or major war was becoming obsolete because they held an explicitly optimistic worldview that even a more competitive, multipolar world would somehow retain the relative stability of the unipolar era and be shaped by the constraining force of globalization…they were channeling Francis Fukuyama, treating historical struggle as finished.[8]

Further, doctrinal doyennes hinder cognitive growth regarding military thinking and thinking about the future of armed conflict by ceaselessly suggesting that formulated institutional military thinking possesses the solution to nearly every military problem – past, present, and future – and in doing so, shackles progressive thought. Doctrinaires hurt much needed critical thinking about the future of armed conflict casting their institutional anchor bias, which keeps military thinking generally moored, regardless of a specific doctrine's successes or failures, or that doctrine's continued relevance or irrelevance regarding the reality of modern warfare.

If this problem is endemic in Western military thinking, as I suggest it is, then what is the solution? The first step in the solution is identifying, naming, and elaborating on the salient problem. I identify the problem as the nexus of strategy, concepts, doctrine, and plans in Western military

---

6 Alexander Montgomery and Amy Nelson, *The Rise of the Futurists: The Perils of Predicting with Futurethink* (Washington, DC: Brookings Institution, 2022), 5.
7 Amos Fox, "The War for the Soul of Military Thought: Futurists, Traditionalists, Institutionalists, and Conflict Realists," *Association of the United States Army*, 17 March 2023, available at: https://www.ausa.org/publications/war-soul-military-thought-futurists-traditionalists-institutionalists-and-conflict.
8 Patrick Porter, "Out of the Shadows: Ukraine and the Shock of Non-Hybrid War," *Journal of Global Security Studies*, 8(3), (2023): 4.

thinking, or what I have referred to elsewhere as the tetrarch of Western military thinking. The tetrarch bakes institutional bias into military thinking, while preventing the injection of new thinking on war and warfare that does not align with its values, preferences, or procurement plans.

Moreover, this approach casts aside matters of reason, logic, and reality for emotion, preference, and ideology. This is where the true problem lies – realism in the study and advancement of military thinking is losing in the race to the future of armed conflict with ideology.

To correct this problem, I posit that the reality of armed conflict is universal and transcends institutional bias and preference, and a result, Western militaries must broaden their understanding of war and warfare to better account for the reality of war. Moreover, thinking about the future of war must be shrewd and penetrating. Thinking about the future of war must see beyond situational fads, faux novelty, and military myths, like many of the supposed lessons of the Nagorno-Karabakh War, or the tank-is-dead bilge which surfaced in response to Ukraine's stalwart defense against the Russian army's mounted offensive in early-to-mid 2022.[9] Instead, military thinking must be critical, it must depend on empirically supportable evidence, and it must make people uncomfortable. To that end, military theory must be better represented in Western military thinking. Balancing military theory against the institutional tetrarch of military thinking will help break the tetrarch's strangle hold on contemporary and future warfare, allowing for a better flow of ideas throughout, and across, Western military institutions.

That is the purpose of this book – to provide a deep injection of realism into the study of conflict. This branch of study I refer to as conflict realism, because it is neither international relations realism, nor any other academic realism. Rather it is oriented on ensuring that realism – that is, what does truth, logic, informed theory, reality, and probability say about the conduct of war and warfare. In applying a Conflict Realist lens to the theory and practice of war and warfare I come to very important findings. First, war and warfare differ. War is the realm of policy and strategy. Warfare, on the other hand, operates from where strategy and tactics meet to the most finite tactical action. As a result of this differentiation, the use of each's word is purposeful and meant to indicate the level of war to which it corresponds.

---

9 John Antal, "Seven Battlefield Disrupters: Warfighting Challenges for the US Military Derived from the Second Nagorno-Karabakh War," *Maneuver Warfighter Conference*, February 2021, available at: https://youtu.be/_At9txsUKIw; Frank Gardner, "Ukraine War: Is the Tank Doomed?," *BBC News*, 7 July 2022, available at: https://www.bbc.com/news/uk-61967180.

In addition, conflict realism operates along a process of iterative analysis of armed conflict. The purpose of this iterative process is to get around observer and analysts bias, and to not operate with possibilistic features of armed conflict, but rather identified probabilistic features of contemporary and future armed conflict. Conflict realism's iterative process moves through three phases. This process strikes to find root causes, causal mechanisms, dependent and independent variables, truth, and applicability.

In Phase I, conflict realists observe an event. The event might be an engagement or a battle, or if looking at something from hindsight, a campaign, or a conflict. Next, the conflict realist applies existing theories, mental models, frameworks, and heuristics to understand the event. During this step, the conflict realist extrapolates findings, trends, linkages, and other nuanced features associated with the event being observed. Next, the conflict realist forms a judgment on what they observed, coupled with their findings. They then use this to state a position. The conflict realists stated position might be a new theory, an update to an existing theory, a set of principles, or any other number of findings relevant to current and future armed conflict. This is done knowing that fidelity will not necessarily be high, but further iteration with the idea will raise its fidelity.

In Phase II, conflict realists take their previous finding and re-evaluate its validity. Conflict realists do this by examining the finding in historical context, current context, how a military forces' training and education prepare it for the environment and did the one's experience color the Phase I position. During the next step, the conflict realist forms a refined judgment. Before moving to generating a new or refined position, the conflict realist cross-examines their refined judgment. They look to see if institutional bias, institutional norms, narratives, or conventional wisdom are interfering with a clear appreciation for realities that impact the conflict. After that case of judgment analysis, the conflict realist articulates a new or refined position.

In Phase III, the process is the same as in the other two. The goal is another opportunity to refine the information. As noted above, the conflict realist iterates as necessary. They should always strive to find root causes, causal mechanisms, dependent and independent variables, truth, and applicability.

This book progresses as follows. Chapter 1 examines the four basic schools of thought that dominate military thinking today – the Institutionalist, the Futurist, the Traditionalist, and the Realist. Each of those schools of thought plays a particular role in modern and future thinking about armed conflict. Nonetheless, this book examines the Realist – of the conflict realist – as its central character, and conflict realism as the prism through which it analyzes contemporary armed conflict and through which it diligently gazes at future armed conflict.

Conflict realism is built on the assertion that military forces – state or non-state – operate in a coherent, self-interested, and value-seeking manner. Reducing this idea to a simple heuristic finds that military forces – state and non-state – operate according to three principles: systems theory, theory of rational action, and economic decision theory. A state, or non-state actors, animates those principles through military power. At the same time, military forces use their power to deny an adversarial force the ability to employ their own power to gain military objectives.

Conflict realism is a theory of war and warfare built on the long-standing position of international relations theorists that states operate to advance their own self-interest.[10] In this situation, states use power to advance themselves along a path from their current position to their desired object, regardless of how far down the line it might be. I find Charles Glaser's definition of power quite useful in thinking about how states – and non-state actors – use power to advance toward their objectives. Glaser asserts that power is a relational concept.[11] That is, power is measured in the context of a specific situation and regarding the other competing actors in that situation.[12] Moreover, Glaser asserts that a state's power is the ratio of its resources that it can transfer into military capabilities in relation to an adversary's ability to do the same.[13]

John Mearsheimer also offers an important point on power. Power, not ideas, hierarchies, or security mechanisms within the international system, determines who gets their way and who writes the rules amongst groups.[14] This idea applies not only at the international level, but is germane in both war and warfare. To be sure, Mearsheimer states as much writing that:

---

10 John Mearscheimer, *The Tragedy of Great Power Politics* (New York: W. W. Norton & Company, 2001), 40-41.
11 Charles Glaser, *Rational Theory of International Politics: The Logic of Competition and Cooperation* (Princeton, NJ: Princeton University Press, 2010) 41.
12 Glaser, *Rational Theory of International Politics*, 41-42.
13 Glaser, *Rational Theory of International Politics*, 41-42.
14 John Mearsheimer, *The Great Illusion: Liberal Dreams and International Realities* (New Haven, CT: Yale University Press, 2015), 16-17.

> Realist logic also applies to other realms besides
> international politics. It goes a long way toward
> explaining behavior in any situation where there is a
> danger that actors will use violence against each other,
> and there is no higher authority to impose order and
> provide protection.[15]

Therefore, one should view power as a critical determinant in both war and warfare. As such, theories of war and warfare should place power at the heart of their ideas. The theory of Conflict Realism does just this.

Chapter 2, Conflict Realism, illustrates a unique school of thought – conflict realism – and a unique theorist – the conflict realist – into discussions of war and warfare. The conflict realist shuns methods of war and warfare that are focused on soft power approaches, or idealistic narratives to explain the inner workings with security partners. The goal of introducing this concept is to reinvigorate thinking within the defense and security studies and international relations communities, and hopefully bring a more genuine appreciation of war and warfare into the fore.

Chapter 3, The Paradoxes of Modern (and Future) Armed Conflict, examines a set of five paradoxes that conflict realism determines to be undermining effective thinking about modern and future armed conflict. I assert that these paradoxes are inhibiting the cognitive maturation of Western military thought. As a result, these paradoxes need to be identified, addressed, and ultimately discarded. These paradoxes include (1) the belief that modern Western militaries have in the primacy of command and so-called "great captains," (2) the current belief that Western militaries place in that small, light, and dispersed forces will overcome the challenges of transparent battlefields and improvement reconnaissance-strike complexes, (3) that battle field transparency will prevent some sort of paralysis of operational and tactical mobility on future battlefields, (4) that Western militaries preference for how they want to fight (for example, their preferred method of warfare) outweighs the way in which the combination in which an opponent's operations, the terrain, international humanitarian law, and other battlefield variables will work together to unhinge their preference, (5) and that defeat mechanisms are still a useful tool for addressing military challenges rooted in systems theory and information-based problems.

---

15 Mearsheimer, *The Great Illusion*, 135.

Chapter 3 continues by providing a set of alternative solutions to the paradoxes outlined above. By focusing on root causes and causal mechanisms, we find that "great captains" and the emphasis on commanders, command posts, and command posts is illogical in modern war because of the role of systems, networks, and information provide in modern (and future) armed conflict. At the most basic level, any army – organized, disorganized, state, or non-state – can operate without a commander, command post, or command note. However, a commander, command post, or command note can do next to nothing without an army (or force) in the field. Thus, in modern (and future) wars, forces are more important than so-called "great captains" or any other element of command.

Further, Chapter 3 makes the case that the current trend in thinking to advocate for small, light forces operating dispersed on future battlefields to overcome transparency, long-range fires, precision strike, and drones is an emotional response to the problem, short-sided, and benefits the opponent in this situation. A more appropriate response is to field large, resilient, and ruggedized land forces that will not culminate short of, or at a military objective, but that possess the capacity to exploit opportunities following initial military victories.

Chapter 3 also argues that one's warfighting preference does not matter. Too many variables are at work to make a military force's preference for something like maneuver at the expense of a pragmatic approach to solving immediate military problem anything more than navel gazing. Thus militaries must adopted concepts and doctrines that address a range of problems and environments to provide policymakers with military forces capable of successfully addressing a range of challenges.

Chapter 4, The Principles and Inverse Principles of War and Warfare, seeks to overturn approximately 100 years of calcified Western military thought. The principles of war, first written by J.F.C. Fuller in the early twentieth century, have remained almost unchanged in Western military doctrine for almost 100 years. Approximately a decade again, when the United States (US) military was struggling to address the myriad of challenges it faced in Afghanistan and Iraq, it added three principles to the long-standing principles of war and changed the name to the "principles of joint operations." The addition of three terms, a true demonstration of almost zero analytical rigor being placed behind what is truly meant by principles of war or principles of joint operations, highlights the lack of serosity paid to the subject in the post-9/11 period of Western military thinking.

Chapter 4 attempts to pick up the principles of war's mantle where it was left by J.F.C. Fuller and put it back at its proper position of respectability. This chapter addresses this problem by first acknowledging that principles are not just one-sided, but they present both what Combatant A wants to do and what Combatant A wants to prevent or deny Combatant B from doing. Put another way, this chapter presents the principles from a positive gain and a negative gains perspective, whereas Fuller and most others just focused on the positive gains position. In stating the principles this way, each principle is presented as a principle (for example, what it wants) and what it hopes to prevent the other from doing or from preventing the enemy combatant from doing.

Moreover, Chapter 4 seeks to improve upon the extant principles of war by decidedly splitting the principles into principles of war and principles of warfare – the former being relative to the law of war at the strategic level, whereas the latter should guide operational and tactical thinking.

Lastly, Chapter 4 attempts to return the principles to the rightful context. When J.F.C. Fuller first began experimenting with the ideas that became the principles of war, they were not written just as a bulletized list, or as a mnemonic for ease of memorization. The principles of war were not a list of words, but a short narrative. Fuller's account of the principles of war was a narrative from which the principles were extracted, as highlights, to help the reader quickly grasp how the elements all worked together as part of a greater whole. That aspect of the elements of war has been lost over time and it is an idea that I reintroduce in Chapter 4.

Chapter 5, On Urban Warfare, explores the realities of urban war. Many notable commenters, to include Anthony King and Jack Watling, clearly illustrate that urban warfare is not only on the rise today but will be increasingly problematic in the future.[16] As military forces continue to decrease in size while cities continue to increase in size, yet the disparity in military capacity and capability between military forces – to include state and non-state elements – continues to grow, urban areas become welcome refuges for less capable military forces.[17]

---

16 Anthony King, "Will Inter-state War Take Place in Cities?," *Journal of Strategic Studies* 45, no. 1 (2022): 74-81. DOI: https://doi.org/10.1080/01402390.2021.1991797; Jack Watling, *The Arms of the Future: Technology and Close Combat in the Twenty-First Century* (London: Bloomsbury Publishing, 2024), 80-92.
17 King, "Will Inter-state War Take Place in Cities?," 74-76; Watling, *The Arms of the Future*, 80.

Chapter 5 explores many of the dynamics that unfold in urban military operations and how states (and non-state actors) operate to both avoid those dynamics, as well as contribute to those problems. In the process of that examination a handful of themes emerged – the pursuit of survival, attrition, sieges, polity and the pursuit of individual victory, the importance of ground lines of communication, and the extreme limitations and shortcomings of precision munitions on modern battlefields. Because of these theme's repetitive character, I made the decision to categorize the most serious and recurrent of them as principles of urban warfare.

Chapter 5 not only provides a set of principles of urban warfare, but like it does with Chapter 4's narratives for the principles of war and the principles of warfare, the principles of urban warfare contain a narrative based problem statement. This problem statement is used to frame the problem that challenges what a military will likely face in an urban environment and how the principles of urban operations can help prepare policymakers, military leaders, and analysts for those challenges.

Chapter 6, titled Sieges and the Consequences of the Urbanization of Warfare, takes a fairly comprehensive look at sieges in the post-Cold War period. As of 2024, my research found that war – which does not include non-war sieges such as the Russia's siege of Beslan of the US' siege of the Branch Davidian Compound in Waco, Texas – in the post-Cold War period has registered 60 sieges. Trends are identifiable when examining the data. Of those 60 sieges, the aggressor won 36, or 60 percent of the sieges. However, when examining time, as sieges move from mid- to long-term (for example, beyond six months and out to a year), an interesting data point illuminate. Defenders become more profitable during the six-to-twelve-month period, winning nearly 58 percent of the time. Nevertheless, beyond a year, the defender's success drops dramatically and would mirror what one might expect to see.

As a result, Chapter 6 provides some insightful analysis into the siege's inherent logic and provides a simple heuristic to graphically explain that logic. Similarly, this chapter provides a detailed description of how resource expenditure of time occurs in an adversarial environment. This heuristic is useful because it can help policymakers, military leaders, and analysts make relatively better-informed decisions about how to approach known or potential sieges prior to their initiation in a specific conflict. Furthermore, this chapter discusses the types of sieges a military force might encounter and how they might best respond to each situation. This is not a set of tactics, but instead a cognitive process that helps a policymaker, military

High Contact– High Movement	High Contact– Low Movement	Low Contact– High Movement	Low Contact– Low Movement
Mobile Warfare	Methodical Warfare	Maneuver Warfare	Entrapment Warfare
Roving Warfare		Positional Warfare	

Table 1.1 Framework for Warfare Matrix

leader, or analyst obtain mental repetitions prior to actually engaging with a true siege in an applied environment.

Chapter 7, On Attrition, is one of this book's boldest chapters. This chapter addresses one of the most lightning-rod topics – attrition and the attrition vs. maneuver debate – in contemporary political, military, and academic debate. In this chapter, I directly address many of the claims made to discredit and undercut attrition's utility on the battlefield, while simultaneously highlighting the absence of empirical rigor behind many of the claims of so many maneuver advocates.

Chapter 7 comes to the conclusion that the maneuver-attrition dichotomy does not exist, nor is it a useful lens through which to frame the argument. To that end, this chapter provides an alternative framework for which to think about warfare. The framework pits contact and movement on competing axes (for example, one on the X-axis and the other on the Y-axis) and weighs each of those variables in terms of 'high' or 'low.' That is, four general types of warfare exist beneath two more broad generalizations of warfare. These generations are based on movement and contact, and therefore, they could easily be re-classified as different terms if the variables used to form the frame's base (see Table 1.1).

Chapter 8, On Precision Strike Strategy, examines precision strike and its place within policy formulation and military strategy in the post-9/11 period. The chapter begins with two small case studies from the US's wars in the Middle East – one in Afghanistan and the other from Iraq. In both cases, the US used the latest precision strike technology, in coordination with the most state-of-the-art intelligence gathering technology, to deliver deadly accurate strikes, both of which proved to be entirely ineffective. The first strike was one from the opening salvo of the US's war against Iraq in 2003 in which it sought to eliminate Saddam Hussain at Dora Farms right before the crisis tipped into armed conflict. The latter was one of many egregious strikes in Afghanistan that hit exactly where and what it intended to strike, but because of incorrect intelligence, it hit an incorrect target.

Chapter 8 then moves on to discuss the trials and tribulations of precision strike in recent conflicts. The chapter uses the Syrian Civil War's battle of Raqqa and Operation Inherent Resolve's battle of Mosul to illustrate the shortcomings of precision strike and precision strike driven military strategy. In both Raqqa and Mosul, the US and its partners employed precision strike handrail in their campaign of annihilation against the Islamic State which resulted in the crude pulverization of each city over the course of many grueling months in relentless combat.

Considering the events, causalities, and connections examined in Chapter 8, a paradox becomes apparent. First, accurate strikes do not necessarily directly correlate into effective strikes. Second, eliminating senior military leadership does not always have an outsized impact on operations in the ways in which military theory supposes. Third, precision warfare often increases, not decreases, civilian casualties and collateral damage. And finally, precision warfare contributes to, not removes, wars of attrition from the international system. The chapter concludes by suggesting that policymakers, practitioners, and scholars be cautious of putting a hire degree of faith into precision strike as a silver-bullet solution to many of the challenges of war and warfare. Instead, policymakers, practitioners, and scholars must evaluate each situation in its own context and make decisions most appropriate to that situation. In some cases, precision might not be the best solution, and therefore, we must examine and invest in other alternatives to precision methods.

The book concludes with a set of recommendations for policymakers, military practitioners and scholars. Conflict realism asserts that the land domain remains the dominant domain in which wars are fought. Other domains, while important, facilitate operations on the land. Therefore, policymakers and practitioners must make funding and procurement decisions that are aligned with this fundamental principle of war. Forgetting that wars are fought and won on the land, by land forces, can result in policymakers and senior military leaders making poor decisions on how to spend money on future force design, weapons procurement, and other areas of DOTMLPF-P (doctrine, organization, training, materiel, leadership and education, personnel, facilities-policy). Conflict realism asserts states must build militaries capable of accomplishing inherent challenges that are fundamental to wars fought on land. Large, ruggedized, and mobile land forces should be the centerpiece of these force designs. The degree of robotization is a discussion for another day, but a discussion that needs to be had after rigorous rounds of experimentation on the subject.

Similarly, conflict realism finds that future conflicts will continue to see states use proxy strategies in new ways that side-step proxy war approaches of both the Cold War and post-9/11 periods. In their place, a newer approach to proxy war is taking shape. This new proxy war strategy accepts the power of information age technology and its ability to all but remove plausible deniability from the battlefield and international system. As a result, states are now exercising proxy strategies that use proxies not to hide their involvement and to generate deniability, but to offload combat to another actor, augment their combat forces, or for a number of other reasons. As such, information age technology is contributing to a flurry of new, and open, proxy wars across the globe. As a result, Western policymakers, military leaders, and scholars must move beyond viewing the term "proxy" as a pejorative and accept the term as what it truly is – just an actor that does work that another actor might otherwise do itself. In accepting the term proxy, Western policymakers and militaries must begin to develop DOTMLPF-P solutions that can help account for proxy strategy in future armed conflict. Concepts and doctrine are the first place to start. Leadership and education should be right on the heels of concepts and doctrine. Moreover, concessions should also be made in force structure to account for proxy strategy. Existing forces might be too engrained in extant modes of thinking about working with partnered forces to embrace proxy war concepts and doctrine, so it might be worth exploring if proxy-type organizations are a suitable solution to some of the future challenges of proxy war problems.

Conflict realism also finds that urban warfare, sieges, and wars of attrition will continue to be significant players on future battlefields. So yet again, we go back to DOTMLPF-P. Western militaries are slowly coming around to the realities that urban warfare, sieges, and wars of attrition present, but not enough is being done. We must push the boundaries of what is acceptable as it relates to military thought, force design, education, and creating formations capable of accomplishing the military operations that policy makers will bestow upon them.

Sticking our heads in the sand because we do not care for certain terms or phrases – whether it is proxy, siege, or attrition – does us no good when we find our forces committed to these very situations. We must do better, think better, debate better, and prepare better for the future of armed conflict. This book, and the idea of conflict realism in general is my attempt to help light the spark of debate.

1 A Taxonomy of Western Military Thought

This chapter sets the book's foundation and does so by exploring the cognitive box Western militaries operate in and how that limits their ability to adequately explore and address problems of present armed conflict. These limitations are not limited to contemporary armed conflict but will also likely persist in the future if not properly accounted for. Furthermore, it provides the theoretical foundation for the idea of conflict realism and then illustrates how proxy wars, urban warfare, sieges, and precision strategy all carry significant weight within the idea.

This chapter proceeds in the following way. First, I examine the institutional tetrarchy of strategy, concepts, doctrine, and plans which maintain agency over military thought and how that structure creates problems for seeing, developing, integrating, and normalizing avant-garde ideas to influence military thought.

Second, I examine how non-institutionally generated military theory can breakdown the established tetrarch's cognitive barriers and push into the intellectual void that exists beyond the bounds of accepted, conventional thought. I argue that military theory, developed by interest practitioners, civilians, and scholars, provides a unique opportunity for thought exploration because it is often unbound by existing military biases, funding cycles, procurement strategies, or other products of institution which might prove as a governor to good theoretical thought.

Third, I posit that contemporary theory is overcome with Futurist ideology, which in and of itself is not necessarily a bad thing. However, considering that most Futurist thought is presented in narrative or axiomatic form, it lacks many of the specifications, principles, and considerations that actually prove useful for institutions seeking novel solutions to future challenges.

Fourth, I provide a basic set of principles for theory development. Conceptually, theory and theory development are universal ideas. Yet, as terms, theory and theory development have different meanings based on the field in which they are applied. To be sure, the articulation of theory and the illustration of theory development in international relations per se, has an accepted method of practice, whereas in military thought this is not the case. Nevertheless, my intent in introducing a set of principles of theory development is to help provide would-be theorists with a set of tools to help make their ideas more marketable in the competitive space of ideas.

I conclude the section on military thought by offering that Western militaries would be well served investing not just in the institutional tetrarch, but by encouraging and rewarding military theorists whether in their midst or those outside of uniform. Budding theorists should not be seen as a threat to institutions, but as individuals of promise who can help militaries avoid dogmatism and remain pragmatic. These theorists, if in the ranks, must be protected from thin-skinned ideologues who feel that questioning the institution is disloyal, and an attack on their self-identify. These theorists should be rewarded and promoted for their contributions to military thought and institutional improvement, and not castigated for potentially non-conformist views. If the theorists reside outside of military institutions, they must be welcomed adjuncts to the organizational tetrarch of military thought.

The Institutional Tetrarch of Military Thought

In general, contemporary Western military thought emerges from four institutional, interrelated subject areas, which include (1) strategy, (2) concepts, (3) doctrine, and (4) plans. Another way to think about these subjects is how they address military challenges. Strategy accounts for military priorities. Concepts incorporate the science of military thought, addressing how to operate, organize, and equip from an experimentation-informed analytical perspective, while adhering to institutional procurement stratagems. Military doctrine, in most instances, builds upon antecedent doctrine, and is rarely innovative. Doctrine carries forth what an institution believes to work, and when updated, generally injects only incremental adaptation. Military plans reflect the synthesis of strategy, concepts, and doctrine, into one of two states a) potential and b) applied. Plans are generally where the art of military thought is reflected – commanders, and more often their staff, apply judgment, experience, and situational understanding through the prism of strategy, concepts, and doctrine to develop a course, or courses, of action ready for the realities of combat.

16   Conflict Realism

Strategy

Strategy is the first pillar of military thought's institutional tetrarch. Scholar Hew Strachan posits that, "Strategy is about war and its conduct...strategy is designed to make war useable by the state, so that it can, if need be, use force to fulfill its political objectives."[1] Accordingly, Strachan states that strategy helps a state define, shape, and understand war.[2] States and their respective militaries draft strategies to address fundamental challenges of state, nascent military conundrums, and ongoing geopolitical necessities.

Most common definitions of strategy define it as a balanced approach to link ends, ways, means, and risk in pursuit of one's prioritized aims.[3] The US military posits that strategies are tools – they are a synchronized group of ideas to leverage the instruments of national power to obtain objectives.[4] Nonetheless, Jeffrey Meiser correctly points out that the US military's *Means-Ways-Ends-Risk* heuristic places resource allocation ahead of innovative thought about how to address political-military problems.[5] Meiser cautions that, "The American way of strategy is the practice of means-based planning: avoid critical and creative thinking and instead focus on aligning resources with goals."[6] Meiser suggests that strategic thought should instead focus on finding solutions to problems by explaining how political-military obstacles can be overcome.[7] To account for this action, Meiser suggests that strategy not be understood through the *Means-Ways-Ends-Risk* heuristic, but rather as a theory of success that is firmly grounded in causal analysis and *Ways* based thinking.[8]

Meiser's misgivings with contemporary thought regarding the purpose and process of strategy are important. Meiser highlights that currently thinking on strategy is generally devoid of innovative and pragmatic thinking, looking to address causal mechanisms, instead focusing mathematically and attempting to solve the problem through the addition, or subtraction, of resources of what Prussian military theorist Carl von Clausewitz referred to as *war by algebra*.[9]

---

1 Hew Strachan, "The Lost Meaning of Strategy," *Survival* Vol. 47, no. 3 (2005): 48-49.
2 Strachan, "The Lost Meaning of Strategy," 48.
3 Joint Publication 5-0, *Joint Planning* (Washington DC, Government Printing Office, 2020): I-3.
4 Joint Planning, I-3.
5 Jeffrey Meiser, "Are Our Strategic Models Flawed? Ends + Ways + Means = (Bad) Strategy," *Parameters* Vol. 46, no. 4 (2016): 82.
6 Meiser, "Are Our Strategic Models Flawed?," 82.
7 Meiser, "Are Our Strategic Models Flawed?," 90.
8 Meiser, "Are Our Strategic Models Flawed?," 86.
9 Carl von Clausewitz, *On War* (Princeton: Princeton University Press, 1986, 141.

Having established what strategies do, it is important to emphasize that the realm of contemporary strategy rarely dabbles with introducing new ideas, and instead stands on tried methods. Perhaps coincidentally, few institutionally ordained strategists, or strategic institutions today generate or contribute to strategic theories about armed conflict, but rather rehash vogue strategic thinking to address nearly all problems. As a result, institutional strategy and institutional strategics cannot be looked to for innovative ideas to address the challenges of future armed conflict.

Concepts – Military and General

Concepts are important because they serve as the mainspring for how forces could operate, equip, and organize for the future. Concepts can also serve as the basis for experimentation and inform DOTMLPF-P (doctrine, organization, training, material, leadership and education, personnel, facilities, and policy) requirements determination, placing them central to how militaries generate material and non-material requirements. Further, concepts inform future science and technology (S&T) priority investments and research.

Concepts pursue the future from what Montgomery and Nelson refer to as a possibilistic perspective. Concepts look at the future with aspiration and attempt to address problems associated with how to fight, what forces and combinations of forces are needed in the future, and what tools those forces need to thrive in the future operating environment. In that pursuit, the concept development process identifies the *what* and the *how* that forces require in the future to achieve their military objectives.[10] These things generally align with emerging material and non-material requirements and both existing and evolving technology.

Concepts also serve as the basis for military experimentation. Emerging concepts are rigorously tested before moving from a nascent idea to a conditionally accepted concept. Wargames, tabletop exercises, and workshop-style idea exploration are several tools for experimentation.

Though synonyms in common English, concepts and theory are not synonyms in Western military thought. As a result, they should not be conflated with one another. Theory is esoteric and, in most cases, it is not rooted in tangible constraints, nor is it governed by the feedback from

---

10 Don Starry, "To Change an Army," *Military Review* 43, no. 3 (March 1983): 25-26.

experimentation in the same way as a military concept.11 In short, concepts are short-hand ideas to link investment and procurement imperatives with operational and tactical warfighting. Concepts can be innovative but only when organizational leaders are willing to push the boundaries of institutional thinking, bias, and pushback.

Doctrine

Doctrine describes the current, procedural aspect of how Army forces fight on modern battlefields. US Army doctrine, for example, states "US Army doctrine is about the conduct of operations...the professional body of knowledge that guides how Soldiers perform tasks related to the Army's role."[12] Doctrine, by virtue of its orientation on executing existing processes in pursuit of accomplishing a mission at hand, resides on a different plane than strategies or concepts.

Like the erosive effect of water on a rock, doctrinal changes in Western military thought are incremental, often coming in drips and drops over the course of many decades. Comparing *AirLand Battle* (ALB) and *Multidomain Operations* (MDO), for instance, finds small conceptual differences between the two doctrines. Both doctrines' focus on joint force-integrated combined arms warfare and winning decisive battles against Russia as their central premise.[13] Both doctrines advocate the use of technology and long-range fires as a central component to their respective theories of victory. Both ALB and MDO place maneuver warfare cognitive warfighting anchor point. ALB's focus on separating the first and second echelon of Soviet forces and winning the first battle therein, offers little difference to MDO's insistence on penetrating an adversary's protective measures to allow ground forces to conduct maneuver warfare and potentially exploit the subsequent tactical or operational success.[14] To be sure, aside for accounting for the technological advancement during the roughly 25 years between the two doctrines' publication, they might well be facsimiles of one another.

---

11 James Rosenau, "Thinking Theory Thoroughly," in *The Scientific Study of Foreign Policy*, ed. James Rosenau and Mary Durfee (New York: Routledge, 2018), 34.
12 Army Doctrine Publication 1-01, *Doctrine Primer* (Washington, DC: Government Printing Office, 2019):
13 Field Manual 3-0, *Operations* (Washington, DC: Government Printing Office, 2022), 4. Field Manual 100-5, *Operations* (Washington, DC: Government Printing Office, 1986), 8-18.
14 Andrew Feicerkt, "Defense Primer: Army Multidomain Operations (MDO)," *Congressional Research Service*, IF11409, 21 November 2022, available at: sgp.fas.org/crs/natsec/IF11409.pdf.

Arguably, doctrine's incremental change reflects a variety of conditions. Doctrinal change can be slow, or marginal evolution because its respective institution, or institutions, are not interested in change. Further, doctrine advancement can be challenged by the individuals charged with developing doctrine. If, for instance, if Institutionalists work on doctrine, idea development will largely orbit around extant organization's thinking.

Additionally, consensus-seeking kills needed doctrinal growth. While a doctrine development team might possess the brightest minds and have developed a set of cutting-edge ideas, external staffing, in which external agencies often seek to protect their interests and advance their own prerogatives, processes can quickly and severely sand away the incisive ideas of novel doctrine thinking, and result in banal, incremental changes.

Organizational leaders who lack foresight or who are set in the ways in which they did things when they were coming up in an institution can also plague needed doctrinal growth. The centrality of maneuver warfare, despite the realities that positional and destruction-based warfighting methods play in modern warfare, is perhaps the most germane example of this idea. Influential military thinkers such as Jack Watling, Michael Kofman, Franz-Stefan Gady, and Anthony King, to name a few, commonly assert that destruction-based warfare, positional warfare, and the relevance of urban operations are germane, and dominant forces in modern armed conflict.[15] Yet, Western military doctrine is slow to adapt and still attempts finds a maneuver warfare solution for almost every military problem, despite how much the situation calls for something else.[16]

One should not therefore look to doctrine for cognitive growth in the face of the potential changes in the future of armed conflict. As the marginal returns between ALB and MDO suggest, doctrine generally keeps the proverbial ship steady, the rudders properly aligned, and militaries oriented on staying the course, while paying lip-service to accounting for the realities of contemporary armed conflict, and the extrapolation of those realities in the future of armed conflict.

---

15 Jack Watling and Nick Reynolds, *Meatgrinder: Russian Tactics in the Second Year of Its Invasion of Ukraine* (London: Royal United Services Institute, 2023); Michael Kofman and Franz-Stefan Gady, "Ukraine's Strategy of Attrition," *Survival* Vol 65, no. 2 (2023); Anthony King, "Will Inter-State War Take Place in Cities?," *Journal of Strategic Studies*, Vol. 45, no. 1 (2022); Amos Fox, "On Sieges," *RUSI Journal*, Vol. 166, no. 2 (2021); Amos Fox, "Reframing Proxy War Thinking: Temporal Advantage, Strategic Flexibility, and Attrition," *Georgetown Security Studies Review*, Vol. 11, no. 1 (2023).
16 Amos Fox, "Maneuver is Dead? Understanding the Components and Conditions of Warfighting," *RUSI Journal*, Vol. 166, no. 6-7 (2021): 10-18.

Plans

Plans are an extrapolation of policy and strategy and the expression of doctrine.[17] By that, plans are the detailed approach to achieve all or part of a strategy, or as US joint doctrine states, "Plans translate the broad intent provided by a strategy into operations."[18] Although plans and concepts both describe how a military force could operate, plans are detailed and intended for implementation and execution.[19]

Plans most often exist to address immediate or emerging problems. As a result, plans provide little-to-no room for the infusion of novel ideas for how to operate, organize, or delineate a battlefield. Plans, therefore, are often reflections of their institution's respective doctrine and their cultural norms, their organization's leaders, and the planners who toil away to create those plans.

What's Missing? The Importance of Military Theory to Western Militaries

Having reviewed Western military thought's tetrarch, we find that relatively little space exists for the exploration of ideas within institutional frameworks. Non-institutional military idea exploration and articulation falls into the category of military theory. Some less strategically minded individuals might be intimidated by, or disinterested in theory, but nonetheless, non-institutional military theory has historically contributed exponentially to the advancement of military thought. Provided that theory manifests outside the confines of headquarters buildings, theory can be the true vehicle for innovative military thought.

Some of the most influential, and lasting, ideas on war and warfare are military theory, and not the result of an institutional tetrarch's rote process. Clausewitz's *On War* – perhaps the most sacred text on political-military thought, was independently published by his wife, Marie, after his death in 1831. Antoin Jomini's *The Art of War* – the United States Military Academy's de facto military instruction manual in the nineteenth century, and rumored to be carried in the pockets of many Civil War general officers – was published after Jomini had hung up his uniform. British theorists J.F.C. Fuller and B.H.

---

17 Army Doctrine Publication 5-0, *The Operations Process* (Washington, DC: Government Printing Office, 2019): 2-1.
18 Joint Planning, I-2.
19 The Operations Process, 2-1.

Liddell Hart published many groundbreaking theories while in uniform, albeit through civilian publishers, and not representing official British Army opinions. Both Fuller and Liddell Hart continued to dominate twentieth century military discourse – official and unofficial – long after retiring from service. In fact, most Western militaries still rely on the principles of war that J.F.C. Fuller developed in 1926 to help guide their wartime activities.

Theory, especially probabilistic theory, began to subside in the mid-to-late twentieth century as Western militaries began to implement more formal control over military thinking through the establishment of centers and commands preoccupied with strategy, concepts, doctrine, and plans. Yet a small coterie of theorists was able to still rise above the power of institutional weight and make a mark on military thought – both institutional and non-institutional – during this period. John Boyd, of OODA (observe, orient, decide, act) Loop fame, and John Warden, and his Five Rings Theory, made significant impacts on institutional military thought right about the time that 1991's Gulf War was underway – both theories proving key elements of the US's war strategy. Robert Leonhard, who published a string of truly impactful works of theory in the post-Gulf War era, also left an indelible mark on military thought, as his presence at contemporary wargames, Western military conferences, and demand as a speaker across Western militaries attests. As Fred Kaplan records, John Nagl and David Kilcullen were important theorists to emerge during the post-9/11 wars as the US and its allies and partners grappled with how to address the political-military problems in Afghanistan and Iraq.[20]

An important question that needs answered is, 'what is military theory'? Strategist Joseph Gattuso posits that theory is, "The track upon which the train runs."[21] By that Gattuso asserts that theory is the intellectual fountainhead that feeds the stream of ideas from which militaries cull important ideas about how to operate and organize for future armed conflict. Further, Gattuso states that theory helps establish methods of operation, which actively contributes to doctrine development, from which nearly all other aspects of military activity follow.[22] Moreover, in stressing the importance of theory to Western militaries, Gattuso emphatically asserts that theory is, "Fundamental to every aspect of the military profession."[23]

---
20 See Fred Kaplan, *The Insurgents: David Petraeus and the Plot to Change the American Way of War* (New York: Simon and Schuster, 2013).
21 Joseph Gattuso, "Warfare Theory," *Naval War College Review*, Vol. 49, no. 4 (1996): 112.
22 Gattuso, "Warfare Theory," 113.
23 Gattuso, "Warfare Theory," 113.

Joel Watson writes that theory is helpful for three fundamental reasons. First, theory provides a language through which to discuss ideas.[24] Further, theory provides the opportunity to construct new conceptual models outside the bounds of dogmatic institutional processes, which supports clear and rigorous thinking.[25] Lastly, theory provides a tool to trace the logical consequences of the assumptions made throughout the process of theory construction; put another way, good theoretical processes allow the theorist to check their work as they go.[26]

Perhaps no military-minded thinker is better suited to answer the question of theory than Clausewitz. He writes that, "The primary purpose of any theory is to clarify concepts and ideas that have become, as it were, confused and entangled."[27] More importantly, Clausewitz illustrates the importance of military theory by stating:

> Theory will have fulfilled its main task when it is used to analyze the constituent elements of war, to distinguish precisely what at first sight seems fused, to explain in full the properties of the means employed and to show their probable effects, to define clearly the nature the ends in view, and to illuminate all phases of warfare in a thorough critical inquiry. Theory then becomes a guide to anyone who wants to learn about war from books.[28]

Additionally, theory is useful because it is versatile. Theory can be linked to hard science, or using Montgomery and Nelson's taxonomy, theory can be probabilistic.[29] In this case, military theory address conflict through reality, logic, reason, and rationality while innovatively thinking about contemporary and future military process, organization, and battlefield delineation. This is often the space in which conflict realists operate. On the other hand, military theory can be completely unhinged from reality, logic, and rationality to paint vivid pictures of less likely futures. This type of theory, in which Futurists often reside, generally falls within Montgomery and Nelson's possibilistic thinking.[30]

---

24 Joel Watson, *Strategy: An Introduction to Game Theory* (London: W.W. Norton and Company, 2010), 1.
25 Watson, *Strategy*, 1.
26 Watson, *Strategy*, 1.
27 Clausewitz, *On War*, 132.
28 Clausewitz, *On War*, 141.
29 Montgomery and Nelson, *The Rise of the Futurists*, 5.
30 Montgomery and Nelson, *The Rise of the Futurists*, 5.

Unlocking the Potential of Military Theory

Considering that most Western military concepts are possibilistic, military theory therefore best serves military thought by adhering to a probabilistic process. Doing so provides ballast between the art and science of military thought and keeping both planes working in unison with one another. Failure to do so can result in outlandish ideas that sound good in lecture halls and conference rooms but fail to deliver lasting impact on the battlefield. Harlan Ullman and James Wade's *Rapid Dominance* theory is a classic example of this idea.

In the mid-1990s, Ullman and Wade wrote that American information, economic, and capability evolutions allowed the US to rapidly attack and overwhelm an adversary's will and information space with precision strike, long-range fires, and speed, and "…produce immediate paralysis of both the national state and its armed forces."[31] Ullman and Wade's theory pushed all the Pentagon's buttons and the concept rapidly moved from an independent theory and to unofficial joint doctrine.

Thereafter, Ullman and Wade's ideas were structural pillars that underpinned the US's invasion of Iraq, in which the war's initial phase was stylized as *Shock and Awe*. Nevertheless, Ullman and Wade's possibilistic theory relied on dubious logic, to include the US's ability to achieve dominant battlefield awareness and perfect (or near perfect) information, both of which are probabilistically unlikely, despite technological overmatch.[32] *Shock and Awe* – as an applied theory – did attain a quick victory against Saddam Hussein's regime, but they also accelerated Iraq's decent into chaos after Saddam's fall, and the subsequent military fiasco. Ironically, Ullman pinned a piece in *The Hill* on the twentieth anniversary of the start of Operation Iraqi Freedom, defending his theory, stating that the theory was sound but that the US military failed to use it correctly.[33]

With Ullman and Wade's story as a cautionary tale, the importance of probabilistic theory becomes apparent. Considering that most Western military concepts are possibilistic, and deeply enmeshed with the ideas of technophile *Futurists*, independent theorists should lean toward developing

---

31 Harlan Ullman and James Wade, *Shock and Awe: Achieving Rapid Dominance* (Washington, DC: National Defense University, 1996), 1-10.
32 Ullman and Wade, *Shock and Awe*, xx.
33 Harlan Ullman, "20 Years On, 'Shock and Awe' Remains Relevant," *The Hill*, 30 March 2023, available at: https://thehill.com/opinion/national-security/3907622-20-years-on-shock-and-awe-remains-relevant/.

probabilistic ideas. A few fundamental ideas are important to probabilistic military theory. First, probabilistic military theory should be grounded in the belief of state-centric power dynamics, and that states are self-interested structures within the international system.[34] This is a foundational idea in military theory because it serves as the point of deviation upon which probabilistic and possibilistic theories develop. Probabilistic theories are generally the result of an actor pursuing self-interest, maximizing investment, operating according to causality, and attempting to best capitalize on the situation at hand. Possibilistic theories, on the other hand, tend to be open less focused on self-interest, which correspondingly means that they operate more on aspiration, ambition, and exploring the realm of the possible. This dichotomy drives the second fundamental idea.

Second, because of probabilistic theory's dependency on a state's self-interested power dynamics, probabilistic military theory also depends on the economic theory of rational choice. Martin Hollis writes that the economic theory of rational choice means that states, or sub-states, operate in ways advantageous for themselves, always seeking to maximize their payoffs.[35] In keeping with this line of logic, states (or sub-state actors) must be assumed to be rational actors. As rational actors, states (or sub-states) use rationality, oriented on cause and effect – or causality – to make decisions. Given the perceived rationality of a so-called rational actor, then the theorist must assume that the actor will make economic advantageous decisions for themselves, first and foremost. Advancing the causality of this idea another step forward finds that rationality is not a conclusive event. Rather, rationality is an iterative cognitive process that involves closely examining known variables, making assumptions regarding unknown variables, and pressing forward with a specific activity. This process is known as sequential rationality.

Third, probabilistic theory should focus on sequential rationality. Sequential rationality is the optimization of a set of ideas, or moves, to maximize the associated payoff of those activities.[36] Aside from a small number of instances, sequential rationality is conditional; meaning each event must be considered based on its own conditions, for example, the known variables and assumptions regarding unknown variables. In

---

[34] John Mearsheimer, *The Tragedy of Great Power Politics* (New York: W.W. Norton and Company, 2001), 11-13.
[35] Martin Hollis, *The Philosophy of Social Science: An Introduction* (Cambridge: Cambridge University Press, 1994), 116.
[36] Watson, *Strategy*, 168.

addition, conditionality is predicated on the fact that entropy affects all things, to include information, and therefore, each effort to optimize to maximize a payoff must be examined regarding the information available at that time.[37]

Fourth, probabilistic theory must appreciate conditional dominance. Conditional dominance is the idea that activities, strategies, or configurations are inherently destined to fail, or be dominated by the adversary, if not conditionally applicable.[38] Fuller, for instance, cautions that mechanized formations operating against light forces in a densely wooded tactical operating environment are conditionally dominated because armored vehicles cannot operate in wooded conditions, while light forces retain their mobility (albeit in a reduced capacity), which facilitates the light forces continued striking activity.[39] Moreover, a force resolved to conduct maneuver warfare will be extremely surprised when they find an adversary elected to not meet them on the approaches to a significant urban area, but instead withdrew into the suffocating confines of the city. The adversary has changed the conditions, requiring the force to either conduct a positional operation to lure the adversary out of the urban area, or they will have to conduct a linear, front assault in which destruction is the currency of victory. Likewise, light and irregular forces can be said to be conditionally dominated if operating against mechanized forces in open terrain, such as deserts of relatively flat plains. This is why battles like 1993's Battle of Mogadishu, Operation Iraqi Freedom's Second Battle of Fallujah, and Operation Inherent Resolve's Battle of Mosul occur. This also accounts for why so many battles in recent wars have occurred in cities – a force realizes that meeting its adversary in another tactical location would result in conditional dominance, and therefore, the force seeks to operate in a non-dominant environment, or at least one that provides them with a fighting chance at victory. Resultantly, probabilistic theory must focus on a force's components (for example, the forces, combat systems, weapon systems, and so on) and the conditions (for example, the adversary and their combat capability, the battlefield's geography, enabling action,

---

37 Watson, *Strategy*, 168; Erik Sherman, "Everything Dies, Including Information," *MIT Technology Review*, 26 October 2022, available at: https://www-technologyreview-com.cdn.ampproject.org/c/s/www.technologyreview.com/2022/10/26/1061308/death-of-information-digitization/amp/.
38 Roger Myerson, *Game Theory: Analysis of Conflict* (Cambridge: Harvard University Press, 1991), 88-89.
39 J.F.C. Fuller, *Armored Warfare* (Harrisburg, PA: Military Service Publishing Company, 1943), 11-13.

and so on) to account for the ephemeral and iterative nature of conditional dominance.

Lastly, probabilistic theory must use backward induction to eliminate, as best as possible, dominated strategies when they arise in theory development. Backward induction and eliminating dominated strategies are important because the probability of a theory's success increases as its propensity to failure (for example, dominated) decreases. In the context of theory development, backward induction is the process of reviewing a theory from the end to the beginning, to identify any logic traps or misalignment of components or conditions that might result in a dominated strategy. During that review process, the theorist must iteratively eliminate dominated strategies.

Closing Thoughts on Theory

To get the most out of military theory, the theorist must proceed with unvarnished judgment regarding interplay among actors in their theories. Theorists must gaze forward in time to consider how a belligerent might respond to various theoretical musings, and vice versa. In theoretical situations in which defeat appears imminent or likely, the theorist should abrogate that move's ideas and explore other options.

Clausewitz provides a further guide for those participants involved in developing military theory. Like latter theorists, Clausewitz states that if the theorist stumbles upon principles and rules in their cognitive travails, all the better. He writes:

> If the theorist's studies automatically result in principles and rules, and if truth spontaneously crystallizes into these forms, theory will not resist this natural tendency of mind. On the contrary, where the arch of truth culminates in such a keystone, this tendency will be underlined.[40]

Considering Clausewitz's guidance, then it is important to understand the theory development is about positive change – for example, adapting to circumstance, environments, technology, and cultural and international norms in beneficials ways to help deliver military victory in the most efficient, ethical, and lasting way possible. In short, military theory is about change, because almost nothing in international relations, military

---

40 Clausewitz, *On War*, 141.

capability, and a state vigorously pursuing their own self-interest is static. As a result, strategies, concepts, doctrines, plans, and all the underpinning ideas therein, are subject to review, evolution, and if need be, discarding.

James Rosenau, on the other hand, provides nine basic principles for good theory and for how to think theory systematically. Rosenau posits that to think theory thoroughly one must:

1. Avoid treating the task as that of formulating an appropriate definition of theory,
2. Be clear as to whether one aspires to empirical theory (for example, how things are) or value theory (for example, how things should be),
3. Assume human affairs are founded on an underlying order,
4. Be predisposed to ask about every event, every situation, or every observation, "Of what is it an instance?",
5. Be ready to appreciate and accept the need to sacrifice detailed descriptions for broad observations,
6. Be tolerant of ambiguity, concerned about probabilities, and distrustful of absolutes,
7. Be playing about the subject,
8. Be generally interested in the subject,
9. Be constantly ready to be proven wrong.[41]

Probabilistic reasoning suggests that five contemporary trends in war will continue to typify armed conflict for decades to come. Those trends include wars of attrition, proxy war, urban warfare, sieges, and the Precision Paradox. These five trends form the keystones in conflict realism's network of ideas. Each of these topics is discussed in varying degrees of detail in scholarly and professional circles. Nonetheless, each of these subjects is explored in greater detail in this book to add a conflict realist's edge to the theoretical debate about those topics.

Conclusion

This chapter serves as a guide for understanding the basic theoretical premise of conflict realism. This chapter asserts that an institutional tetrarch maintains a strong hold of contemporary military thought, which

---

[41] James Rosenau, "Thinking Theory Thoroughly," *The Scientific Study of Foreign Policy* (London: Frances Pinter, 1980), 19-31.

is stifling the connection of independent ideas with the Western military thought. Although a few works of independent theory, like August Cole and P.W. Singer's *Ghost Fleet* and *Burn In*, or Elliot Ackerman and James Stavridis' *2034*, have been able to impact institutional military thought in recent years, this is the exception and not the norm. Interestingly, the works that tend to make it into the tetrarch of Western military thought tend to be fictional novels and not academically written works of theory. Perhaps that is a lesson for modern theorists to carry forward.

Further, I provided a formula for theory development to serve as a companion to the institutional tetrarchs of Western military thought, which are often possibilistic in their outlook. I assert that budding theorists should focus on creating probabilistic theory to counterbalance institutional possibilistic strategies, concepts, doctrine, and plans. Further, I posit that five considerations are germane features of probabilistic theory. First, probabilistic theory must be firmly set in the idea that states, and sub-state actors, are self-interested and value-seeking participants during armed conflict. Second, owing to the fact that most states, and sub-states, operate according to the economic theory of rational choice, then probabilistic theory should be approached with the assumption that all participants in war and warfare are rational actors. Third, pursuant to the assumption that all actors act rationally, theorists must approach theory development through sequential rationality. Fourth, theory development should ruthlessly eliminate dominated strategies through the application of conditional dominance. Fifth, probabilistic theory must use backward induction to doublecheck for dominated strategies.

Conflict realism is one of the primary schools of thought in military thought. Conflict realism is a theory of war and warfare. Moreover, conflict realism is a theory of war and warfare that is the product of probabilistic military theory based. Conflict realism is not aspiration or built on hope, but instead looks for the causal mechanisms in armed conflict and conducts analysis and forecasts based on the belief in the importance of causality. Conflict realism is explained in detail in Chapter 2.

2 Conflict Realism

This chapter provides the theoretical foundation for the idea of conflict realism. Conflict realism rests on two foundational pillars. First, conflict realism rests on the belief that technology cannot, nor does not, overcome the human variables in war and warfare. The human variables are best understood as the impact of terrain, that of enemy interactions on the battlefield – to include their capabilities and intentions, and time on military activities. Whereas futurist, technologist, and institutional thinking tries to upend the influence that the human elements have on war and warfare, conflict realism asserts that those factors are a deterministic force on military forces that are engaged armed conflict. Second, conflict realism attempts to surmount the cultural chauvinism that permeates military thought to lay bare the realities of armed conflict.

What's more, conflict realism is a subordinate element of realist international relations theory. Conflict realism adheres to the assertion that the state is the main unit of measure in the international system. But, because conflict realism is focused on military thought and not international relations scholarship, it uses military forces as its unit of measure. Further, conflict realism believes in the idea that power is the currency of international relations, and consequently of armed conflict.

To support these ideas, this chapter proceeds as follows. First, I briefly examine how the human elements of war and warfare interact result in combatants operating toward primary objectives – survival and winning. Second, I illustrate how conflict realism fits as an offshoot theory of realist international relations theory. I examine how cultural chauvinism undermines clearly understanding causality in war and warfare. From there, we frame conflict realism, illustrating how the theory fits in the space between international relations and tactical military thought, but relatively parallel to – but distinct from – policy and military strategy. From there, this chapter moves to illustrate the linkages between conflict realism and international relations realism. This chapter then concludes with a set of truths and verities pertaining to conflict realism.

Conflict Realism: Rediscovering the Deterministic Impact of Human Variables

In the past, military thought examined the application of forces in relation to terrain, one's enemy – real or theorized – and time. Today, military thought examines the application of technology, largely devoid of the impact that terrain, an enemy (their capabilities and intentions), and time have on military operations. These elements – terrain, an enemy, and time – can be thought of as the human variables in war and warfare. The obviation of the human variables in war and warfare is likely the result of technologists, futurists, and Western military institutional bias toward believing that technology obviates or explicit surmounts the human variables of war and warfare.

Today's focus on technology in war and warfare, absent the human variables, creates flawed strategies, concepts, doctrines, and plans. Moreover, it creates unrealistic policy objectives. As James Rogers writes, the idea that the coupling of precision strike and persistent information collection capabilities might somewhere make war less bloody or dangerous is a popular myth that has existed as long as militaries have possessed precision strike capabilities.[1] For instance, much of the leading scholarship about precision strike today fails to make the simple adjustment of an adversary's basic desire to survive. A combatant's desire to survive can be accounted for by simply assuming that an adversary will not meet their opponent on terrain favorable to the aggressor, nor in ways that enhance their inherent capabilities. The Chechens use of urban terrain, anti-armor weapon systems, and quick-hitting light infantry tactics against the vaunted Russian land forces during the First Chechen War's Battle of Grozny is an example of how the human elements in war often outweigh, or at least counterbalance technological superiority.[2] Moreover, a similar dynamic unfolded when Russian forces invaded Ukraine in February 2022. Ukrainian forces – both regular and irregular – used similar technology and very similar tactics as the Chechens to stop Russia's attacks on Kyiv and Kharkiv dead in their tracks within days of the invasion.[3] At the time of the invasion, Russia the

---

[1] James Rogers, *Precision: A History of American Warfare* (Manchester, England: Manchester University Press, 2022), 8-18.
[2] Stasys Knezys and Romanas Sedlickas, *The War in Chechnya* (College Station, TX: Texas A&M University Press, 1999), 90-113.
[3] Mark Galeotti, *Putin's Wars: From Chechnya to Ukraine* (Oxford: Osprey Publishing, 2022), 346-348.

columnists and the international community bestowed Russia with near boogeyman status because of the Kremlin's perceived military capability and capacity to wage region wars.[4] In 2016, for instance, US think tank RAND published a report stating that the Russian military could entirely overwhelm and take control of the Baltic states within 60 hours, if Putin and the Kremlin so choose to invade their westerly neighbors.[5] These ideas, and supporting case studies, are examined in greater detail later in this book. Nonetheless, the lacuna in military thought that these examples point to have grown dangerously large since the technology and information systems booms of the 1990s.

Conflict Realism: Confronting Cultural Chauvinism in Military Thought

If the examples used in the preceding paragraph caused you to take pause and say something to the effect that: "The Russian military is bad, it has always been bad, and it will always be bad, and thus, we – whomever that might be in your situation – would never make those mistakes," then you are part of the second problem that conflict realism attempts to address. The second problem is one of cultural chauvinism.

Cultural chauvinism is the belief that one's own military force, or any specific force, is better than another just because of who they are. For instance, during the US Civil War, Confederates were apt to say that one confederate soldier was as good as five Union soldiers.[6] Likewise, during World War II, the Germans viewed nearly every other culture and race of people as inferior to their own, and thus, the German soldier was inherently better than any other state's soldiers.[7]

Conflict realism is an unfettered approach to conflict analysis that strips away cultural chauvinism. By doing so, conflict realism attempts to see though the layers of bias that cultural chauvinism creates to find the truth of a matter, regardless of how popular or unpopular that truth

---

4 Molly McKew, "The Gerasimov Doctrine: It's Russia's New Chaos Theory of Political Warfare and It's Probably Being Used on You," *Politico*, 5 September 2017, accessed 4 January 2024, available at: https://www.politico.com/magazine/story/2017/09/05/gerasimov-doctrine-russia-foreign-policy-215538/.
5 David Shlapak and Michael Johnson, *Reinforcing Deterrence on NATO's Eastern Flank* (Santa Monica, CA: RAND Corporation, 2016), 4.
6 Cathal Nolan, *The Allure of Battle: A History of How Wars are Won and Lost* (Oxford: Oxford University Press, 2017) 430.
7 Nolan, *The Allure of Battle*, 432.

might be. Take the precision paradox, for instance. The precision paradox is covered in a chapter later in this book. However, the precision paradox is basically the finding that precision strikes, although often accurate, are often ineffective, meaning that they were unsuccessful in achieving their objective. The US often uses self-righteous language when describing how it uses precision strike. In press releases, for example, the US military often makes specific mention of using precision munitions, instead of standard ballistic munitions.[8] At the same time, the US military will criticize Russia for not using precision munitions.[9] Yet, the US military's use of precision strike to effectively level cities like Mosul in 2016-2017 and Marawi in 2017 is left unsaid when advocating the US's precision strike prowess and denigrating other combatants, such as Russia, for their poor use of precision strike. Conflict realism attempts to set aside the cultural chauvinism that serve as hurdles to making a clear-eyed assessment of political and military situations.

The theory of concept realism thus seeks to bring the human variables in war and warfare back into focus. The goal is to accomplish this while stripping cultural chauvinism away from political and military analysis. The goal of which is to help decision-makers develop more appropriate and unbiased policies, strategies, concepts, doctrines, and plans. The tertiary goal is that conflict realism provides a methodology to help decision-makers see beyond the narratives and information operations meant to soften the realities of war and warfare, and therefore make decisions that better account for war and warfare's deadly and destructive nature.

Framing Conflict Realism

Conflict realism is an idea created to fill a void in scholarship and military thought. At the international level, international relations and international

---

8 "Statement from Secretary of Defense Lloyd J. Austin III on US Strikes in Iraq," *Department of Defense*, 25 December 2023, accessed 5 January 2023, available at: https://www.defense.gov/News/Releases/Release/Article/3626180/statement-from-secretary-of-defense-lloyd-j-austin-iii-on-us-strikes-in-iraq/.
9 Phil Stewart, "Exclusive: US Assesses Up to 60% Failure Rate for Some Russian Missiles, Officials Say," *Reuters*, 25 March 2022, accessed 5 January 2024, available at: https://www.reuters.com/business/aerospace-defense/exclusive-us-assesses-up-60-failure-rate-some-russian-missiles-officials-say-2022-03-24/; Paul Adams, "Ukraine War: Why Russia's Infrastructure Strikes Strategy Isn't Working," *BBC News*, 9 March 2023, accessed 8 January 2024, available at: https://www.bbc.com/news/world-europe-64900098.

affairs scholarship exists to analyze and debate the interaction amongst states. International relations scholarship witnessed a boom in the period following World War II and now the field of study is rich with a panoply of leading scholars both old and new and it is the home of well-established theories and concepts.

In international relations scholarship, the state is generally the primary unit of measure. International relations scholarship also examines why states fight, the causes of international armed conflict, the frequency of war and events within war, and how war ends. To a smaller degree, international relations scholarship looks at how states fight, that is, what are the strategies that states employ when engaged in conflict and armed conflict. Insurgency, counterinsurgency, and proxy war scholarship are some of the more common examples of the 'how' subset of international relations scholarship. Nonetheless, very little of the discussion on 'how' states engage in conflict within international relations scholarship addresses military forces.

States use military strategy to link policy and the 'how' subset of international relations. States use military doctrine to educate, train, and guide their military forces. However, a rather large gulf exists in the space between international relations and how state's use military forces in a practical sense. Conflict realism is one attempt to solve this problem and fill this cognitive vacuum with a general theory. Within conflict realism, military forces – state, non-state, and otherwise – take the place of the state as the unit of measure. I refer to this changing of the analytical lens as a subject-based assessment, or SBA.

The SBA uses the subject, the target, the methods, and the intentions as the variables to examine a particular situation. The subject is clear enough – the object under review of the unit of measure used to examine a particular situation. In the case of conflict realism, a military force, instead of a state, is the most common subject. Since both states and non-state actors possess military forces, I use the term combatant most often used to identify a subject, or subjects, throughout this book. A threat is the competing actor against whom a subject focuses its political and military activity. Within the school of conflict realism, the subject and the threat are military forces, unless otherwise stated. The methods are a subject's strategy for applying the principles of war and principles of warfare in pursuit of political and military victory. Resources and imagination are the limiting factors for a military force's methods.

Scholarly Foundations

Conflict realism can be seen as a derivative of the realist school of international relations theory. First and foremost, conflict realism adheres to Thomas Schelling's assertion that, "War is a process of violent bargaining."[10] Further, conflict realism conforms to Schelling's statement that, "Each party's strategy is guided mainly by what he expects the other to accept or insist on; yet, each knows that the other is guided by reciprocal thoughts."[11] Conflict realism's orientation on the ideas of survival, victory, and the challenge-response cycle are where Schelling's ideas find a home.

Kenneth Waltz, Alexander Wendt, and other leading international relations scholars contend that states are the unit of measure for examining the international system.[12] To be sure, Waltz asserts that, "States are the units whose interactions form the structure of the international-political systems," and that "So long as the major states are the major actors, the structure of the international system is defined by them."[13]

Conflict realism is a theory of international relations that takes a similar outlook to Waltz and likeminded realists. Yet, instead of focusing on how states operate within the international system and the reasons for the how, why, and reasons states make war, conflict realism seeks to understand how military forces – both state and non-state – conduct war. This level of analysis lies just below the line of international relations scholarship and as a companion to conflict and security studies.

Moreover, most realists adhere to the belief that, "The behavior of the great powers is influenced mainly by their external environment, and not by their internal characteristics. The structure of the international system, which all states must deal with, largely shapes their foreign policies."[14] Conflict realism shares this belief and finds that the same logic applies to military forces, when they replace the state as the unit of measure and armed conflict replaces international politics. Put simply, conflict realism

---

10 Thomas Schelling, *Arms and Influence* (New Haven, CT: Yale University Publishing, 1966), 23.
11 Thomas Schelling, *The Strategy of Conflict* (Cambridge, MA: Harvard University Press, 1980), 70.
12 Kenneth Waltz, *Theory of International Politics* (Long Grove, Illinois: Waveland Press, Inc., 2010), 94-95; Alexander Wendt, *Social Theory of International Politics* (Cambridge: Cambridge University Press, 1999), 198-199; John Mearsheimer, *The Tragedy of Great Power Politics* (New York: W. W. Norton & Company, 2001), 17; Charles Glaser, *Rational Theory of International Politics: The Logic of Competition and Cooperation* (Princeton, NJ: Princeton University Press, 2010), 29.
13 Waltz, *Theory of International Politics*, 94-95.
14 Mearsheimer, *The Tragedy of Great Power Politics*, 17.

believes that the behavior of military forces involved in armed conflict is influenced mainly by their external environment – to include the capabilities and intentions of competing military forces – and not by their internal characteristics.

Power is the third key position that realist scholars find important to appreciate international relations and this is no different for conflict realists. John Mearsheimer writes that, "Realists hold that calculations about power dominate states' thinking, and that states compete for power among themselves."[15] Further, he writes that power is "Nothing more than specific assets or materiel resources that are available to a state."[16] Charles Glaser writes similarly that, "The strategies that are available to a state and their prospects for success depend on the opportunities and constraints created by the state's international environment."[17] Writing about applied power, Robert Dahl famously says that Actor A possesses power over Actor B insofar as Actor A can make Actor B do something it would not otherwise do.[18] Moreover, Dahl discusses the material factors required to exercise power, stating that Actor A's power is limited to the depth and breadth of their bases of power – real, active, latent, and emerging.[19] He continues, stating that whichever power can mobilize and maintain material overmatch is likely to exert dominant power over the other actor.[20]

Conflict realists operate according to a similar belief structure regarding power, albeit one focused on military forces in armed conflict, and not states operating in the international system. Conflict realists accordingly believe that military forces resources cap the extent to which a military force can achieve its political, strategic, operational, and tactical level goals. Limited resources and a limited base of power, then the limited power one possesses and the limited power it can exert to try and attain its objectives. Conflict realism believes that power and bases of power, however, are not anchored on initial starting conditions. A military force can experience a growth or decrease in material and applied power over the course of conflict as allies, coalitions, and partners come and go. Therefore, initial estimates about what a military force can and cannot do in an armed conflict should be tempered by the belief that all the major sides will generate more power

---

15 Mearsheimer, *The Tragedy of Great Power Politics*, 17.
16 Mearsheimer, *The Tragedy of Great Power Politics*, 26.
17 Glaser, *Rational Theory of International Politics*, 33.
18 Robert Dahl, "The Concept of Power," *Behavioral Science* Vol. 2, no. 3 (1957): 202-203. Doi: 10.1002/bs.3830020303.
19 Dahl, "The Concept of Power," 203.
20 Dahl, "The Concept of Power," 203.

in the opening phases of conflict. This is because military forces – and their respective governments – will contact friends and potential friends to negotiate joint war plans or, as Paul Poast classifies the process, to build military alliances.[21] Poast states that these joint war plans are limited by the degree to which the potential allies possess compatible war plans and how attractive outside offers are to the incumbent partner.[22] As a result, conflict realists believe that the military balance prior to and during the initial phases of a conflict do not directly correlate to a war's outcome. Take the US-led war in Afghanistan, which lasted for 20 years. By all material measures, the US, NATO, and other partners should have easily prevailed in the conflict. Yet, the US did not appropriately account for the structure of the conflict – to include the direct and indirectly involved participants – that allowed the Taliban to abscond from political power, rebuild military power, and successfully retake control of the country when the US withdrew in disgrace from the country in August 2021.[23] In Iraq, the US and its partners faced a similar dynamic. The US withdrew from Iraq in December 2011, after eight long years of war. In departing, the US left victory on Iran's doorstep and Iranian-aligned politicians controlling large portions of the Government of Iraq.[24] In a few short years, this created the conditions that fueled the rise of the Islamic State, who quickly and easily wrested control of significant parts of western, central, and northern Iraq from the Government of Iraq. US material power and applied power mattered little in the fight for Iraq because the US failed to understand the structure of the conflict's system in sufficient time to make meaningful adjustments to compensate for that disparity. Thus, policymakers, strategists, and military practitioners must be reasonable about the limits to which their military forces can achieve and not assign them objectives beyond the realm of reality.

Moreover, conflict realism adheres to the realist principle that, "Land power is the dominant form of military power in the modern world."[25] Other types of warfare – air, sea, cyber, and so on – are factors that must be considered when examining the balance between states and military

---

21 Paul Poast, *Arguing About Alliances: The Art of Agreement in Military-Pact Negotiations* (Ithaca, NY: Cornell University Press, 2019), 14.
22 Poast, *Arguing About Alliances*, 14-15.
23 "U.S. Withdrawal from Afghanistan," *White House*, 1 April 2023, accessed 12 January 2024, available at: https://www.whitehouse.gov/wp-content/uploads/2023/04/US-Withdrawal-from-Afghanistan.pdf.
24 Jeanne F. Godfroy, et al., *US Army in the Iraq War, Volume 2: Surge and Withdrawal* (Carlisle Barracks, PA: US Army War College Press, 2019), 576-598.
25 Mearsheimer, *The Tragedy of Great Power Politics*, 83.

forces, but land power is the true indicator of a combatant's power. This is increasingly important when non-state actors are added to the discussion.

Most non-state actors do not possess air or sea power, but that does not prevent them from possessing considerable power and influence with a specific conflict. In the US-Afghan War (2001-2021), for instance, the Taliban make that point undeniably clear. The Taliban lacked air or sea power but were able to: a) push the US into a 20-year stalemate, b) force the US into giving up the Afghan government and Afghan security forces, c) compel the US to withdrawal from the country, and d) return to power. Moreover, the Islamic State in Iraq is another example of this dynamic. Despite using small drones for reconnaissance and discrete attacks, the Islamic State did not possess an air force or naval capabilities. Nonetheless, they were able to prove a significant challenge for the Iraqi Security Forces, the US, and the US-led international coalition to militarily defeat. Although the blistering siege of Mosul is the best-known battle from the counter-Islamic State conflict, a handful of smaller, but no less bloody and destructive battles preceded Mosul, with smaller sweeps following Mosul. In all, it took the Iraqi security forces, the US, and the US-led coalition – who collectively, dominated the land, sea, and sky – more than three years to defeat the Islamic State in Iraq.

Further, states and their territorial boundaries are fundamentally rooted in the earth, and so long as men and women are willing to fight and die for a line that separates two states, land warfare will remain the main martial consideration in armed conflict. Moreover, US dominance in the air and sea did not deliver victory in Afghanistan, nor did it fuel quick victory over the Islamic State. As Mearsheimer reminds the reader, "Wars are won by big battles, not by armadas in the air or on the sea. The strongest power is the state with the strongest army."[26] Although, considering the discussion on non-state actors, Mearsheimer's assertion can be amended to read that the strongest power is the state, or non-state actor, with the strongest or most resilient land force. Air forces, navies, space forces, and cyber forces are all important in their own right, but their true importance is their respective contribution to unlocking combined arms and joint operations for military commanders. For these reasons, land forces and land operations are the primary method through which military forces are used to examine war and warfare within this book.

---

26 Mearsheimer, *The Tragedy of Great Power Politics*, 84.

Truths of Conflict Realism

Within the bounds of conflict realism, a series of truths exist to govern the analytical space. These truths are not constructs of imagination but are the product of a detailed literature review on war and warfare. The truths are grouped into how wars are won, how wars are lost, and verities of armed conflict.

Wars are Won By

1. Political acquiescence wins wars; credible and capable military forces are the tools that facilitate political acquiescence.[27]
2. Materiel endurance wins wars.[28]
3. Human endurance wins wars.[29]
4. Combatants win wars by iteratively cycling through destructive campaigns that deplete an adversary's political materiel, and human capital over time.[30]
5. Combatants win wars by applying pragmatic, situationally relevant solutions to unique problems on a battlefield.[31]

Wars are Lost By

1. Combatants lose wars by fighting for policy objectives that are beyond the reach of their resource capability or military capability.[32]
2. Combatants lose wars by having policy objectives which are out of step with domestic risk tolerance.[33]
3. Combatants lose wars by being materially weak.[34]
4. Combatants lose wars by trying to win quickly.[35]
5. Combatants lose wars by being dogmatic, idealistic, and mentally

---

[27] Carl von Clausewitz, *On War* (Princeton, NJ: Princeton University Press, 1986), 75-77.
[28] Jurgen Brauer and Hubert Van Tuyll, Castles, Battles, and Bombs: How Economics Explains Military History (Chicago: University of Chicago Press, 2008), 322-324.
[29] Clausewitz, *On War*, 85-87.
[30] Glaser, *Rational Theory of International Politics*, 41.
[31] Geoffrey Blainey, *The Causes of War* (New York: The Free Press, 1988), 173.
[32] Stphen Van Evera, *Causes of War* (Ithaca, NY: Cornell University Press, 1999), 27.
[33] Clausewitz, *On War*, 606.
[34] Glaser, *Rational Theory of International Politics*, 35-36.
[35] Nolan, *The Allure of Battle*, 577.

recalcitrant toward embracing new ideas considering evolving situations.[36]

Conflict Realism's Verities of Armed Conflict

1. An adversary always has more resources and bases of power than what correlation of forces and means (COFMs) comparisons prior to, and during the initial days of a conflict, indicate. Just as COFMs cannot accurately account for the depth of a combatant's resources and range of bases of power, neither can net assessments.[37] Net assessments are one actor's (or analyst's) appraisal of the military balance of a specific subject.[38] Net assessments are based on known information, inferred information, and general assumptions. Because states and their military forces possess a degree of private information, and because publicly stated policy and strategy is often used to advance a state or military force's narrative instead of truthful policy and strategy, a true net assessment is often beyond the bounds of reachability.[39]
    a. COFMs cannot account for known unknowns, such as if an adversary has partners that will help when conflict comes, but do not know for sure who those partners are and what range of support the partner will provide.
    b. Net assessments are incomplete assumptions about how an adversary might operate, organize, and equip for armed conflict.
    c. Suboptimization factors should be applied to any COFMs comparison, net assessment, or associated policy, strategy, or plan to account for the impact of private and incomplete information. The less information available about an adversary, the higher the suboptimization factor.
2. An adversary always has partners who will materialize after a conflict has begun, and thus the adversary's base of power will expand. Bases of power are, according to Robert Dahl, "All the resources – opportunities, acts, objects, etc. – that he can exploit

---

36 Van Evera, *Causes of War*, 27.
37 Glaser, *Rational Theory of International Politics*, 41.
38 Eliot Cohen, "What is Net Assessment?," *Institute for National Security Study* (1990): 4. Doi: https://www.jstor.org/stable/pdf/resrep08961.4.pdf.
39 Van Evera, *The Causes of War*, 45.

in order to effect the behavior of another."[40] This will strengthen the adversary and increase its political and military durability. Moreover, this dynamic will cause wars to be longer than anticipated and accelerate conflicts towards – not away from – an attritional characterization.
   a. COFMs analysis is always misleading.
   b. Conflicts will elongate as more partners materialize.
   c. As conflicts elongate, they will become more attritional.
3. An adversary is always more elastic than what onlookers think, and more resilient than what the analysis suggests. As Stephen Van Evera asserts, "The record indicates that real first-move advantage is rare while the illusion of first-move advantage is common."[41] Therefore:
   a. Few modern combatants are prone being defeated in short and decisive wars.
   b. Centers of gravity are not useful tools for policy, strategy development, or any other forms of planning.
   c. Aside from temporary moral boosts, the importance of success in 'first battles' is not important.
4. Any combatant's own actions are more prone to suboptimization than they realize and therefore anticipate.[42] Carl von Clausewitz refers to this idea as the impact of friction in warfare. Clausewitz writes that, "Countless minor incidents – the kind you can never really foresee – combine to lower the general level of performance, so that one always falls far short of the intended goal."[43] Clausewitz writes, "Friction is the only concept that more of less corresponds to the factors that distinguish real war from war on paper."[44]
   a. Lofty, ethereal, and complex goals should be avoided.
   b. Goals that require the alignment of several variables should be avoided.
   c. Goals should be achievable.
   d. Goals should be tangible, measurable, and observable.

---

[40] Robert Dahl, "The Concept of Power," *Behavioral Science* Vol. 2, no. 3 (1957): 203. Doi: 10.1002/bs.3830020303.
[41] Van Evera, *The Causes of War*, 71.
[42] Van Evera, *The Causes of War*, 25.
[43] Clausewitz, *On War*, 119.
[44] Clausewitz, *On War*, 119.

e. Goals should be reframed episodically to ensure that they are still achievable, tangible, measurable, and observed.
f. All situations in armed conflict are more challenging than they appear. As Clausewitz cautions, "Everything in war is very simple, but the simplest thing is difficult."[45]

Conclusion

A vast amount of literature discusses how and why states go to war. To be sure, this is international relationship scholarship's *raison d'être*. Many think tanks and public policy institutes do address the role of military forces, but in many cases the source of funding can (and does) influence those organizations' analysis and findings. Yet today, the area below the state – that of the military force – is devoid of detailed, and unbiased, analysis. This is a relatively new phenomenon that somewhat parallels the rise of international relations scholarship in the early-to-mid twentieth century. Prior to that time, military theorists such as J.F.C. Fuller, B.H. Liddell Hart, J.C. Wylie, and many others, dominated the discourse of military affairs. Military theorists, unlike international relations theorists, sought to understand how and why military forces operated the way that they did, why recurrent trends in wars and warfare were recurrent, and how to overcome those military challenges.

Although not empirically proven at this point, the rise of international relations as a discipline somewhat coincided with the decline of military theory as a discipline. As a result, the contextual issues of military thought are missing from the discussion of war and warfare today. Conflict realism – the concept, and this book – seeks to fill that void by bringing the military force level of analysis back into the fold.

Conflict realism provides an alternative method to examine war and warfare. Conflict realism is a subset of the realist school of thought from international relations scholarship. As a result, conflict realism adheres to the belief that states are the primary unit of measure within the international system. However, unlike realism and international relations scholarship in general, conflict realism replaces the state as the unit of measure with military forces. The purpose is to move the analytically lens a level below the start. Doing so shifts analysis from why, how, and how frequently states go to war – which is the primary focus of most international relations

---
45 Clausewitz, *On War*, 119.

scholarship – to how military forces and why military forces operate the way that they do.

Shifting the analytical lens from the state to military forces opens the aperture for an enticing new world of insights to flood into view. That enticing new world is what I hope to offer as we continue the journey through this book. Although many topics emerge from shifting the analytical lens from the state to military forces, this book addresses a handful of those topics, to include military strategy, urban warfare, sieges, attrition, precision strike strategy, and proxy wars and proxy war strategy.

This chapter provides a set of truths and verities that should be remembered as the reader works their way through the remainder of this book. The principles and verities are ideas that resonate across the basic framework of conflict realist thought and help advance the principles associated with conflict realism. Lastly, conflict realism seeks to portray war in light of the human variables of war – many of which carry a deterministic effect on military forces, and therefore, military operations – and devoid of cultural chauvinism. Cultural chauvinism – or the thought process that places one military force below another based on who they are, and not as a result of rigorous analysis – causes states to misjudge an adversary, or potential adversary. As a result, cultural chauvinism often causes one combatant to think that the other is inferior and can be quickly defeated with a small, light force, which history shows is almost always mistaken.

The hope is that this book brings the military theorist and military thought back from the grave. While international relations debates the actions of states and militaries discuss how to use their forces on the battlefield, very few individuals or organizations are taking a clear-eyed look at military thought and using military forces as the subject through which to examine war and warfare. Conflict realism – the concept – is an attempt to help breathe life back into the field of military thought.

In Chapter 3, I continue to lay the theoretical foundation for conflict realism by examining a series of paradoxes that dominate discussions of both war and warfare today, as well as potentially in the future. The paradoxes are the product of unquestioned suppositions made by contemporary Western militaries and think tanks to advance various self-interests. Nonetheless, the paradoxes are unhelpful toward the study of armed conflict because they redirect a potential student from understanding and appreciating the truth about armed conflict, and instead push them towards a manufactured reality that is not fundamentally correct.

3 The Paradoxes of Modern (and Future) Armed Conflict

In this chapter, I examine a set of paradoxes that inhibit the cognitive maturation of Western military thought. The neglect of adaptive and self-interested oppositional innovation is the theme that binds each of these paradoxes and why they provide limited utility for the practitioner and scholar of armed conflict.

To counterbalance these paradoxes, this chapter provides a set of standards for each paradox that is grounded in the belief that technological innovation has a quick half-life. This quick decay time is because self-organizing adversaries always respond in self-preserving strategies to battlefield novelty. The purpose of the paradoxes outlined in this chapter are two-fold. First, it seeks to spur a penetrating examination of conventional wisdom by diverging from the tyranny of institutional hive mindedness. This is done by directly examining several ideas within the defense and security studies communities that are today taken at face value. Second, it attempts to provide the intellectual stimulus needed to energize the creation of theories and concepts compatible with the reality of armed conflict and adequate for the future of war and warfare.

As a point of clarity, before beginning on this chapter's exploration of ideas, this chapter is theoretical. Steven Van Evera states that theory, "Is nothing more than a set of connected causal laws or hypotheses."[1] John Mearsheimer defines theory as a simplified description of reality that helps explain how some facet of the world works.[2] Further, Mearsheimer states that theories consist of empirical claims, assumptions, and causal logic.[3] Empirical claims are based on the relationship between independent,

---

[1] Stephen Van Evera, *Guide to Methods for Students of Political Science* (Ithaca, NY: Cornell University Press, 1997), 12.
[2] John Mearsheimer, *How States Think: The Rationality of Foreign Policy* (New Haven, CT: Yale University Press, 2023), 38.
[3] Mearsheimer, *How States Think*, 38.

dependent, and intervening variables.[4] Assumptions are informed but yet unproven assertions about how independent variables impact their dependent counterpart. Stated another way, assumptions offer explanations for causal claims. In the spirit of international relations theory, and the other social sciences, I rely on hypotheses, hypothesis testing through process-tracing, and case studies within this book to examine and advance theories pertinent to military forces.

I define military theory as the analytical space just below international relations theory, in which the object of analysis is the military force, and not the state, or non-state polity, as it is in international relations theory. Yet, military force does not make military theory explicitly a tactical examination of how forces fight one another in engagements and battles. Rather, military theory orients on military force operations primarily just below state-level analysis for how and why states engage in armed conflict. By changing the object from the state to the military force, the level of analysis focuses on the relationship and interplay between policy and military strategy, and military strategy with military operations. In some instances, military theory dips into tactical discussions, but only when additional, low-level context is needed to help illustrate a larger theoretical point.

Today, many defense, security, and conflict studies communities do not see military theory in a positive light. This is the product of the belief that military theory is concocted and not based on analytical rigor. Moreover, many academic fields view theory with an explicit meaning and, as a result, their definition of theory is bound with systematic processes. As Stephen Van Evera writes, for instance, Political Science practices a basic set of methods, of which theory occupies a discrete place and relatively precise set of steps.[5] In conflict studies, however, this practice is not necessarily carried into theory development. Conflict theory might be written with testable hypotheses or conveyed as general conceptual ideas. Despite these notions, this chapter uses qualitative techniques, focused on process tracing and the cause-and-effect of causal mechanisms to support the hypotheses herein. The goal of using these techniques is not to generate an ephemeral theory which waxes philosophical, but a theory that is practical for real world utilization.

Communication challenges can dominate the discourse of ideas when theoretical debate becomes decorated with flourishes, which in the

---

4 Mearsheimer, *How States Think*, 38.
5 See Stephen Van Evera, *Guide to Methods for Students of Political Science* (Ithaca, NY: Cornell University Press, 1997).

case of theory is often pedanticism and the use of impenetrable language. For theory to be useful, the theorist must provide crisp, neatly framed arguments, regardless of the idea's complexity. As a result, the paradoxes and principles outlined within Chapter 4 attempt to address the challenges of war and warfare, in addition to coded institutional bias, head-on and leave the dense language of social science on the cutting room floor.

Further, theory differs from concepts, at least in most Western militaries. Technically speaking, concepts, unlike theories, are generally warfighting ideas that experience some degree of experimentation (for example, wargaming, simulation and so on). As a result, a concept likely possesses more analytical, and perhaps more quantitative rigor than a theory; whereas a theory likely possesses more conceptual detail than a concept and is potentially more qualitatively and empirically informed than a concept. Further, individuals interested in the study and application of war and warfare as a tool of international relations create theories. Concepts, on the other hand, are the product of institutions. As a result, institutional bias leeches into almost every aspect of a military concept. Thus, Western military concepts (and their derivative doctrines) might be less wedded to reality than theories, and more focused on advancing unshakable institutional preferences, procurement strategies, and ideas that resonate with the decision-makers who control financial investment.

Western military thought during the previous 20 years has been heavily focused on using non-linearity and emergent behavior to try and explain the workings of military forces, that include state militaries, non-state actors, and contractual proxies. In this environment of armed conflict, Western military thought elevated precision strike to nearly panacean status. In doing so, Western military thought during the past 20 years has pushed general military strategy from a tool to address holistic problems in armed conflict to something akin to issue management via a targeting process.

Evidence, in the form of a string of failed Western wars in Afghanistan, Iraq, and Libya, among others, indicates that the extant Western military logic outlined above – both in its theoretical and applied sense – has proven insufficient to address the challenges of self-interested actors seeking the polar goals of strategic victory and not being defeated. Emerging Western military ideas, such as Multidomain Operations (MDO), Combined Joint All Domain Command and Control (CJADC2), and Convergence, continue to reflect the 'issue management-via-targeting process' approach to strategy, operations, and doctrine. In effect, the warfighting rationale emerging in recent years further reduces military thought to what one could be described as a 'strategy of point tactics', in which focusing inordinate amounts of precision firepower at a 'decisive' point can prove strategically relevant. The underlying logic behind these ideas remains anchored in the long obsolescent belief that centers of gravity (COGs) are some sort of magic button, and that if a belligerent can just strike the appropriate element within a COG, they can force their adversary into strategic paralysis and subsequent collapse with nary a fight.[6] Nevertheless, Cathal Nolan and Anthony King, among others, note that wars are not won through a falsifiable supposition about the causal relationship between precision strike, cognitive well-being, and decision-making. Instead, they provide irrefutable evidence that contemporary armed conflict – especially when the combatants are industrialized states – continues to portray that war and warfare are attritional affairs in which victory is wrought through strategic exhaustion.[7] The wars between Russia and Ukraine and Azerbaijan and Armenia, as well as the growing unrest in the Balkans, validates many of Nolan and King's assertions. Further, Nolan and King's assertions are of increasing importance as the US and other Western states examine the series of hard challenges that would accompany a potential conflict with China regarding its Taiwanese ambitions.

Today a set of vogue conceptual arguments, categorized as paradoxes within this chapter, corrupt clear-sighted understanding of both war and warfare and thereby reinforce the 'issue management-via targeting process' problem plaguing Western military thought. These paradoxes are (1) the belief that commanders and command nodes bear significant value on modern and future battlefields, (2) that small, light, and more deployable forces are required to face the challenges of future armed conflict, (3) that the future of war will see a more transparent battlefield, (4) that one's warfighting preference matters, and (5) that a defeat mechanism is legitimate, and if so, is actually helpful. These paradoxes, in turn, help provide fodder for revitalizing the principles of war heuristic, which is one of this book's subsequent contribution.

---

6 Franz-Stefan Gady, "What if the Deep Battle Doesn't Matter?," Peter Roberts (host), *This Means War* (podcast), 14 September 2023, accessed 16 September 2023, available at: https://podcasts.apple.com/us/podcast/what-if-the-deep-battle-doesnt-matter/id1629454648?i=1000627835531.
7 See Cathal Nolan, *The Allure of Battle: A History of How Wars Have Been Won and Lost* (Oxford: Oxford University Press, 2017) and Anthony King, *Urban Warfare in the Twenty-First Century* (Cambridge, England: Polity, 2021).

Paradoxes of Modern Military Thought

Paradox 1: The Command Paradox

Great Captains
The conventional wisdom amongst military experts suggests that a linear relationship exists between battlefield victory and command, control, and military leadership.[8] This relationship is as old as war itself. Perhaps no historical example better illustrates the causality associated with the Command Paradox more than the Napoleonic War's War of the Third Coalition. In this conflict, heads of states, and other lesser political dignitaries, often either led their armies into battle, or at least accompanied them on the battlefield.[9] In this political-military configuration, a significant and definitive military victory over an adversary often generated an outsized geopolitical impact. Battle, due to the proximity of the head of state to the death and twisted metal of victory or defeat, was the currency of armed conflict.[10] Further, most armies of this period, and those operating in conscription systems, required significant supervision to prevent desertion. Therefore, eliminating military leaders could cause disorder to overtake conscript-based armies, whereas eliminating the head of state, or at a minimum, eliminating the political leader's army in a battle, could bring that state or polity to heel.

During Austerlitz, the heads of state for France, Austria, and Russia were all present on the battlefield, which resulted in the battle also being referred to as the battle of the Three Emperors.[11] The emperors' presence ensured that no geopolitical dithering or mission creep followed. Austerlitz's outcome – like so many battles in which political leadership physically accompanied their armies in the field – forced the emperors to make definitive, quantifiable decisions about the war's outcomes. The ensuing Treaty of Pressburg encrusted Bonaparte's geopolitical position

---

8 Milford Beagle, Jason Slider, and Matthew Arrol, "The Graveyard of Command Posts: What Chornobaivka Should Teach Us about command and Control in Large Scale Combat Operations," *Military Review*, Vol. 103, no. 3 (2023).
9 Owen Connelly, *Blundering to Glory: Napoleon's Military Campaigns* (Lanham, MD: Rowman and Littlefield Publishers, 2006): 86-89.
10 Amos Fox and Thomas Kopsch, "Moving Beyond Mechanical Metaphors: Debunking the Applicability of Centers of Gravity in 21st Century Warfare," *Strategy Bridge*, 2 June 2017, accessed 24 September 2023, available at: https://thestrategybridge.org/the-bridge/2017/6/2/moving-beyond-mechanical-metaphors-debunking-the-applicability-of-centers-of-gravity-in-21st-century-warfare.
11 David Chandler, *The Campaigns of Napoleon* (London: Scribner, 1973), 413.

(at least temporarily), while leaving no question to Austria and Russia's political and military defeat. In this configuration, a simple linear logic (*Logic 1*) existed: *Leader* → *Will to Resist*.

A leader's presence is not just important to geopolitical decisiveness. Leadership and command nodes are important to militaries built on conscripts and uninspired soldiers. Leadership serves not only as a motivating factor in warfare, but also as a tool to implement and maintain order, where disorder would otherwise exist. In these systems, leaders maintain order over a relatively unmotivated mob so that they can be used as a vector of mass violence against adversaries. Likewise, in these types of systems, leaders serve as the keystone to their respective military's central nervous system. A leader's location, in these systems, are where decisions are made, information and intelligence are digested, and strategy and plans are formed. Moreover, in periods of conflict before decentralized decision-making became chic, all command-and-control decisions came from these leaders. A simple linear logic (*Logic 2*) exists in these armies: *Leader* → *Order* → *Operate*.

Today's militaries are generally professional and purpose-oriented, unlike their conscript counterparts of the past. Modern militaries, in most cases, represent their state and pursue their state's foreign policy, and the associated military objectives. In the past, armies were more representative of a head of state and the local commanders who maintained order and inspired the less dedicated soldiers to accomplish the head of state's military objectives. The 'Great Captain' theory of military leadership is borne out of *Logics 1 and 2*. Despite fundamental changes in the causalities and purpose of command and control, *Logic 1* and *Logic 2* remain dominant in Western military thought. The reasons for this are debatable, but nonetheless, the logic perseveres.

Looking at the problem today, however, militaries represent the state, and resultantly, the state invests in its military. States are not interested in seeing their militaries quickly defeated. Therefore, states build forces that are resilient and that have depth to persevere when engaged in sustained combat. Today, the case should be made that a different logic is truly present in armed conflict. That logic (*Logic 3*) can be summarized in the following heuristic: *Perseverance* → *Order* → *Operate*. Again, mission accomplishment, or victory, is the handrail by which all actors advance, albeit acknowledging that all definitions of victory are both fundamentally state (or actor)-specific, and not entirely known to an adversary, despite the best intelligence effort.

In acknowledging the logic of *Perseverance* → *Order* → *Operate*, it follows that the importance of commanders, headquarters nodes, and command posts loses luster. *Logic 3* systems, of which most Western militaries operate, use a state-centric rationale, unlike the individual-centric rationale of *Logics 1* and *2*. In *Logic 3*, therefore, individual leaders are of less importance to the state than is their military's ability to endure the rigors of battle and continue to mission accomplishment. To be sure, *Logic 3* produces very few General Bonapartes because the systemic rationale that underpins *Logic 3* is built on a 'next leader up' approach, and not individualistic leadership. As a result, the 'Great Captain' theory likewise also falls to the wayside as states emphasize impersonal, organizational leadership.

Logic 3 will become of increasing importance in the future. The diffusion of artificial intelligence, machine learning, and autonomous systems will further erode the importance of individual-minded approaches to command and control. Information, data, data processing, and information networks, not individuals, will replace 'Great Captains'. Human analysis will become decreasingly important as machines, having sifted vast amounts of information and generated a situationally prioritized list of options, relegate commanders to selecting options, authorizing high-risk missions, and owning the aggregate effect of operations within their respective area of operations. If artificial intelligence, machine learning, and autonomous systems assume their supposed position of importance in the future, 'Great Captains' will be replaced by 'Great Networks' and 'Great Authorizers'.

Root causes

Another significant problem with *Logics 1* and *2* is that they do not appropriately address root problems in modern armed conflict. *Logics 1* and *2*, which dominated nineteenth century battlefields, fall apart when an adversary eliminates a force's commander or command elements. This is important to highlight because nineteenth century military logic, captured in the works of Carl von Clausewitz and Antoine Jomini, remains the bedrock of contemporary military thought today. To be sure, a side-by-side comparison of the works of Clausewitz and Jomini with Western military operations doctrine, for instance, provides a strikingly similar set of ideas. The presence of terms like decisive point(s), center of gravity, and interior and exterior lines, among many others, across those works is one such example of how *Logics 1* and *2* rationales remains foundational to modern military thought.

Today, however, states use armed forces that can obtain information, generate targeting data, and strike from increasingly longer distances. On today's (and tomorrow's) battlefield, information, information networks, forces, and strategic purpose are (and will be) more important than individual commanders. Networks and information are challenging, if not impossible, to eliminate today because they do not exist in the physical world. One might even argue that because networks and information are non-physical variables, they cannot be destroyed, but only disrupted. The destruction of a command post or headquarters, for example, does not destroy the information or the network pulsing through that command node. Instead, the strike only temporarily destroys the means through which a small number of humans interact with the information on a network.

Strategic purpose is another non-physical variable. Like networks and information, strategic purpose cannot be destroyed. Because heads of state do not lead their armies into armed conflict any longer, strategic purpose – embodied by the political decision and supporting military objectives to engage in armed conflict – thus remains an ethereal idea that exists in the minds of those operating on the battlefield.

In today's period of armed conflict, in which *Logic 3* pulses through military activity, commanders, headquarters, and command posts no longer carry the significance that they did when *Logics 1* and *2* carried the day. As a result, an adversary's force is truly the most important battlefield variable to strike. Although this perspective is controversial, and likely elicits an emotionally charged feeling by the reader, it is important to examine this hypothesis in more detail.

Commanders, headquarters, and command posts cannot operate without forces capable of accomplishing a mission. A commander with nothing to command is just an individual. Likewise, a headquarters or command post with no forces to direct and support is just a collection of individuals occupying a workspace.

A force in the field without a commander nonetheless possesses the capability, purpose, and information to accomplish its military objectives. A force in the field without a command post or headquarters can operate on strategic purpose, their last orders, or be organized beneath another battle flag. Modern networks ensure that information flow, despite the absence of a commander or a command node, continues to pump throughout a military's digital information networks. The absence of a commander is handled through the 'next leader up' process of pragmatic problem

Variable	Can operate	Cannot operate
A commander without force		●
A headquarters without force		●
A command post without a force	●	
A force without a commander	●	
A force without a headquarters	●	
A force without a command post	●	

Table 3.1 Logic 3: Command to Force Comparison

solving, which is a feature of nearly every modern state military and non-state actor military force. As a result, modern military forces can generally be expected to continue operating toward their military objectives until they no longer possess the physical means, the self-discipline, or they have accomplished their mission. Table 3.1 illustrates the causal effect of *Logic 3* between command and forces.

Considering the causal impact of *Logic 3* on modern military operations, two broad topics can be deduced. First, as a general rule, eliminating commanders, headquarters, and command posts is incongruent with the established and well-worn routes to military victory. Commanders and computers are quickly replaced, but tanks, artillery, and people are not. Systematically annihilating an adversary's forces stops its ability to advance its state toward their policy objectives.[12] Force-oriented military operations, however, amplify an adversary's strain of industry and increasingly complicates an opponent's strategic economic considerations in ways that simply extirpating a commander or command node cannot compete.

Second, a strategy that targets the elements of command through precision strike might heighten, not diminish, long and destructive wars. Considering the causality identified in Table 3.1, forces without commanders and command elements will fight on until they cannot, but a commander without a force can do nothing. The logic follows that the methodical eradication of an adversary's military force from the battlefield will result in victory sooner than the elimination of commanders and command elements. Yet, reality tells us otherwise.

---

12 Jurgen Brauer and Hubert Van Tuyll, *Castles, Battles, and Bombs: How Economics Explains Military History* (Chicago: University of Chicago Press, 2008), 131.

The US military's precision strike campaign that targets al Qaeda and the Islamic State's leadership throughout the Middle East, for instance, has killed commanders almost as the seasons change, yet both organizations maintain engaged forces in armed conflict.[13] Further, military operations in Ukraine generally supports the causality of Paradox 1. Throughout the summer of 2022, Ukraine, with the assistance of US intelligence, methodically eliminated upward of 15 Russian generals on the battlefield.[14] The targeted killing of Russian generals might have disrupted Russian military activities which were already poorly coordinated, resourced, and planned, but the impact on the specific battles, and the war as a whole, were limited at best. Aside from generating a lot of chatter on social media in support of Ukraine, the strikes did little to accelerate Russian military defeat and increase the odds of Ukrainian military victory.

Paradox 2: Small, light, and dispersed is better
Conventional thought amongst Western militaries and many analysts today suggests that small, light forces which are dispersed across a battlefield are best suited to address the challenges of future armed conflict.[15] This suggestion is predicated on a belief in the increasing importance of 'non-contact' warfare, or the idea that a state can sense an adversary with

---

13 Mark Landler, "20 Years On, the War on Terror Grinds Along, With No End in Sight," *New York Times*, 10 September 2021, accessed 30 September 2023, available at: https://www.nytimes.com/2021/09/10/world/europe/war-on-terror-bush-biden-qaeda.html; Jeff Seldin, "Death of Islamic State Leader Not Seen as Diminishing Long Term Threat," *VOA News*, 9 August 2023, accessed 30 September 2023, available at: https://www.voanews.com/a/death-of-islamic-state-leader-not-seen-as-diminishing-long-term-threat/7217471.html.
14 Julian Barnes, Helene Cooper, and Eric Schmitt, "US Intelligence is Helping Ukraine Kill Russian Generals, Officials Say," *New York Times*, 4 May 2022, accessed 11 September 2023, available at: https://www.nytimes.com/2022/05/04/us/politics/russia-generals-killed-ukraine.html; David Martin, "Gen. Mark Milley on Seeing Through the Fog of War in Ukraine," *CBS News*, 10 September 2023, accessed 11 September 2023, available at: https://www.cbsnews.com/news/gen-mark-milley-on-seeing-through-the-fog-of-war-in-ukraine/.
15 Sydney Freedberg, "Army of 2030: Disperse or Die, Network and Live," *Breaking Defense*, 17 October 2022, accessed 11 September 2023, available at: https://breakingdefense.com/2022/10/army-2030-disperse-or-die-network-and-live/; Todd South, "Army Prepares for Dispersed Warfare with High Casualties," *Army Times*, 11 October 2022, accessed 11 September 2023, available at: https://www.armytimes.com/news/2022/10/11/army-prepares-for-dispersed-warfare-with-high-casualties/; Randy Noorman, "The Russian Way of War in Ukraine: A Military Approach to Nine Decades in the Making," *Modern War Institute*, 15 June 2023, accessed 11 September 2023, available at: https://mwi.westpoint.edu/the-russian-way-of-war-in-ukraine-a-military-approach-nine-decades-in-the-making/.

increasing clarity and at amplified distances, and accordingly, strike the adversary from increased ranges with boosted precision.[16]

Further, a large amount of the advocacy for smaller, lighter forces reference videos from the Nagorno-Karabakh War of 2020 and short, sensational video clips from the Russo-Ukrainian War, in which drones and precision strikes liquidate single tanks and small fighting positions.[17] Reflecting on these snippets of finite tactical engagement, many commenters suggest that the lesson of these engagements is that large forces are destined to be identified and destroyed on future battlefields and that drones and precision strike are increasingly the answer to addressing land warfare challenges.[18]

Yet, stringing together disparate clips of discrete strikes and poor tactical acumen hardly represents analytical rigor, and accordingly, it should not form the basis of supposed lessons learned, nor recommendations to force structure changes. Viewing armed conflict through the lens of engagements and targeting processes can cause a state to inappropriately prepare for war, both strategically and tactically. States and their militaries must appreciate that wars are won through laborious and resource extensive campaigns. More strikingly, military victory results from exhausting an adversary's tactical systems and military formations, their associated industrial base, and their commitment to remain immersed in the conflict.[19]

Moreover, the threat of nuclear war, which thread throughout the Cold War provides an analog to the perceived challenges of battlefield transparency and long range-precision strike. Following World War II and the Soviet Union's demonstration of nuclear warfighting capability, the US Army asserted that large and slow formations would be quickly identified

---

16 Raphael Cohen, et al., *The Future of Warfare in 2030: Project Overview and Conclusions* (Santa Monica, CA: RAND Corporation, 2020), 69-77.
17 Shannon Bond, "How Russia is Losing – and Winning – the Information War in Ukraine," *NPR*, 28 February 2023, accessed 29 September 2023, available at: https://www.npr.org/2023/02/28/1159712623/how-russia-is-losing-and-winning-the-information-war-in-ukraine.
18 John Antal, "Learning from Recent Wars – Observations from the Second Nagorno-Karabakh War and the Russian Ukrainian War, *European Security and Defence*, 5 October 2022, accessed 29 September 2023, accessed at: https://euro-sd.com/2022/10/articles/27498/learning-from-recent-wars-observations-from-the-second-nagorno-karabakh-war-and-the-russian-ukrainian-war/; James Hasik, "Precision Weapons Revolution Changes Everything," *CEPA*, 17 February 2023, accessed 29 September 2023, available at: https://cepa.org/article/precision-weapons-revolutionize-russias-war-in-ukraine/.
19 Brauer and Van Tuyll, *Castles, Battles, and Bombs*, 124-127; Russell Weigley, *The Age of Battles: The Quest for Decisive Warfare from Breitenfeld to Waterloo* (Bloomington, IN: Indiana University Press, 1991), 542-543; Nolan, *The Allure of Battle*, 577.

by Soviet reconnaissance and destroyed on the battlefield in tactical nuclear strikes.[20] The Pentomic Division was thus the solution to the theoretical problem of being easily found and quickly destroyed by short and long-range nuclear strikes.

Richard Kedzior asserts that, the Pentomic Division, which intended to use dispersed operations with small, light, and more readily deployable formations to counter the threat that nuclear weapons posed to large, heavy formations, was fraught with paradox when it transitioned from theoretical musing to the reality of field operations.[21] Kedzior states that military leaders failed to account for a handful of important considerations. First, military leaders failed to appreciate the true destructive power of tactical nuclear weapons which, in reality, exceeded the utility of small, light, and dispersed forces. Second, the small, light, and dispersed forces lacked the combat power and lift to deliver useful strikes, and distributed dispositions ladened the ability to quickly mass to attack targets of opportunity, or conduct overlapping defensive operations.[22] Third, the Pentomic Division's small, light, dispersed forces created logistics nightmares for sustainers due to the time in which they were on the road.

The Pentomic Division's small, lighter, dispersed forces were, as R.F.M. Williams records, to rely on "nuclear fire support, dispersion, speed, and mobility." Moreover, a Pentomic Division's subordinate forces required "hyper mobile" forces, a large logistics pools to feed "small, scattered supply points" that sustained forces close to the front. Yet, as Williams correctly highlights, the US Army could not cross the Rubicon and make the Pentomic Division meet the expectations of theoretical supposition regarding dispersed operations.[23]

The issues raised by Kedzior and Williams provide a good starting point to examine the call, yet again, for small, light, and rapidly deployable forces suited for distributed operations. Both Kedzior and Williams' points remain valid today, and likely well into the future. In addition, however, three additional points must be raised. First, distributed operations by small forces will attract more attention, not less, from an adversary. A small force moving into and out of a location is likely more noticeable than a larger

---

20 Richard Kedzior, *Endurance and Evolution: The US Army Division in the Twentieth Century* (Santa Monica, CA: RAND Corporation, 2000), 24-26.
21 Kedzior, *Endurance and Evolution*, 27.
22 Kedzior, *Endurance and Evolution*, 28-29.
23 R.F.M. Williams, "The Rise and Fall of the Pentomic Army," *War on the Rocks*, 25 November 2022, accessed 11 September 2023, available at: https://warontherocks.com/2022/11/the-rise-and-fall-of-the-pentomic-army/.

force operating in the same space. What's more, it will likely be easier, not more challenging, to find important support and enabling elements in small formations than in larger formations because discrete units and capabilities are better hidden within large formations than in smaller ones.

Second, sustainment and logistics activities occurring within a larger formation tend to be less visible than in small formations, for example, sustainment and logistics activities get lost in the background noise of a large formation's continuous movement, striking, and protecting. Similarly, small formation sustainment and logistics activities, absent some kind of internal generation of classes of supply, tend to telegraph the location of sustainment units, tactical warfighting units, headquarters, and the connecting road networks. As a result, distributed operations by smaller, lighter forces provide a watchful adversary with a simpler targeting and reconnaissance process.

Third, small, lighter, and dispersed forces make that force prone to piecemeal destruction. An opportunistic adversary can use a dispersed force's distribution to its own advantage. The adversary can do so by inserting itself between disparate units, blocking reinforcement to one or more of the dispersed units, and eradicate the isolate units, one by one. Nonetheless, two additional challenges present significant hurdles to the necessarily of small, light, more deployable forces and the problem of battlefield transparency.

Challenge-Response Cycle

Militaries operate within a competitive adversarial environment. Within that environment, competing actors are perpetually engaged in a challenge-response cycle. In that cycle, for every nascent solution an actor introduces into the conflict, their adversary will seek a response, and correspondingly introduce a nascent solution to that problem.

Cautious of falling prey to circular logic, it is important to highlight that modern precision munitions, sensing, drones, and long-range strike likely fit within the 'challenge' phase of a renewed period of state-centric, industrialized international armed conflict. It is reasonable, therefore, to assume that second-mover states, and their friends, partners, and allies, are collectively contributing to the 'response' phase of the challenge-response cycle and attempting to develop countermeasures to today's novel warfighting capabilities. By virtue of the time lag that exists between a response's ability to catch, and generate parity, with a challenge, then the logic of linearity generally supposes that the technology and tactics of one actor's 'response' phase remain a step behind the technology and tactics of their adversary's 'challenge' phase.[24]

Today's drones, long-range fires, and GPS and radar innovations are examples of challenges at the fore of industrialized international armed conflict. Certainly, the military thought community is replete with chapters stating that drones and long-ranges fires are, and will continue, to revolutionize warfare. Yet, a large portion of the literature originating from this cognitive space fails to account for a conflict's adversarial context and how the challenge-response cycle uniquely shapes how competing states ebb and flow between advantage, parity, and disadvantage.

Considering the fluctuation between challenges and responses in the context of armed conflict, then, it is reasonable to expect a limited, and short-term impact of any technological innovation in war and warfare. Thus, it is likely incorrect to immediately assume that lighter, smaller forces operating in a more dispersed manner are the answer to nascent technological innovation pertaining to precision strike, battlefield transparency, and enhanced target sensing and identification. Perhaps what is more needed than an impulsive reaction is a heavy degree of intellectual investment in counter-sensor, deception, and counter-rocket and counter-missile technology. Moreover, the investment in those capabilities should possess a tactical and operational component, meaning that when examining how to address those problems, solutions should focus on small unit concepts and technology, as well as larger unit investment. For instance, a massed brigade, that possesses the ability to obfuscate its location and its movement pattern, containing counter-missile and rocket systems, and can repel tactical drones, will arguably be a more useful formation for Western militaries than would a brigade that does not possess those additional capabilities, that is dispersed to the point that physically massing is challenging, that its dispersion limits its ability to mass effects at a range that provides any degree of standoff, and who's distributed array contributes to an adversary's targeting process.

Small, light forces and the challenges of land warfare
The 'light footprint' approach possesses a vampiric quality in military thought. From the failed Pentomic Division, to the 'light footprints'

---

[24] Andrew Carr, "It's About Time: Strategy and Temporal Phenomena," *Journal of Strategic Security*, Vol. 44, no. 3 (2021), 315-316.

inability to seal victory and prevent disaster in both Afghanistan and Iraq, to Russia's flawed invasion strategy of Ukraine in February 2022, the 'light footprint' always appears to be the answer to a future problem of armed conflict yet always fails to deliver on its perceived value. Although this assessment is purely qualitative, a quantitative assessment would likely yield an even more incisive, and likely a more damning, finding.

Military history does illustrate, however, that large formations and large footprints are more adept and better enabled to address the challenges of land warfare. This is not necessarily due to their physical agility or dexterity, but to their ability to provide military commanders flexibility to address a range of challenges, where a small, light force provides a limited flexibility due to its small size.

In practical terms, the US Secretary of Defense Donald Rumsfeld's mandate for a small, light footprint allowed US forces in Afghanistan to quickly topple the Taliban-led Afghan government in October 2001. Yet, the small, light force failed to provide sufficient manpower to prevent senior leaders from the Taliban and Al Qaeda from escaping into the mountains of Afghanistan and into Pakistan.[25] Scholar Steve Coll asserts that the US would have needed at least an additional 2,000 to 3,000 soldiers to block al Qaeda's disappearance into Pakistan.[26] Moreover, the iterative cycle of troop surges in Afghanistan, which often accompanied changes in command of the war effort, also reflect the small, light footprints inability to provide military commanders the tools they need to accomplish their military missions.

US General Tommy Franks' insistence on a light footprint in Iraq provides another example of how the small, light footprint fails military commanders. Likely interested in remaining in favor with Donald Rumsfeld, Franks' plan for Iraq called for an extremely small force – little more than a traditional US Army Corps, plus the required joint force attachments.[27] Buying into information circulated by the Central Intelligence Agency, Franks and his acolytes build their plan on the belief that, following a military victory against the Iraqi military, the US would be welcomed by Iraqis as liberators. As such, there would not be a need for a large military footprint because the US would not be an occupying power, and sovereignty would

---

25 Steve Coll, *Directorate S: The CIA and America's Secret War in Afghanistan and Pakistan* (New York: Penguin Press, 2018), 105.
26 Coll, Directorate S, 105.
27 Michael Gordon and Bernard Trainor, *Cobra II: The Inside Story of the Invasion and Occupation of Iraq* (New York: Vintage Books, 2006), 109-117.

be quickly turned back over to the Iraqis.[28] Nearly none of this happened. Iraq quickly devolved into a massive, country-wide, insurgency which the US military spent the better part of the next eight years trying to overcome. A small, light force posture was not the answer to the land challenges in Iraq. A troop surge and deployment extensions, to increase the land force on the ground in Iraq, were initiated to help the US military keep pace with the rapidly evolving challenges occurring on the ground.[29]

Further, the US military faces a unique dichotomy among most state militaries. The US military is expeditionary – neither of its bordering states are hostile and it possesses almost no regional military challenges – which causes the US military to be deployed globally. By virtue of seeking self-interest across the globe, a small, light military is preferable because it can muster quickly, board transport, and be on the ground on the opposite side of the globe in relatively short order. This matters both for contingency operations and larger military operations, like the invasions of Afghanistan and Iraq.

Although the ability to swiftly move across the globe requires rapidly deployable forces, land warfare, and the many challenges therein, tends to require large forces that can handle a myriad of tests that exceed the bounds of rapid deployability.[30] In many cases, those challenges – such as urban warfare, control of terrain, or human-to-human engagement – cannot be overcome with long-range firepower, stand-off strike, or any other number of technological innovations. Instead, those challenges require individual soldiers, and their leaders, on the ground, making timely decisions that require human judgment, in the moment, and in the context of the situation, to prevail.

Further, by virtue of being expeditionary, US military forces must possess sufficient weight of force to conduct forcible entry. This can include anything from sea-borne land embarkation to airborne assaults. Because of the high margin for error associated with these operations, especially if one factors in a hostile adversary intent on preventing the success of US expeditionary operations and forcible entry, the margin for error, regardless of technological innovation, also increases. Therefore, it must be assumed that larger, more capable land force, not smaller and lighter land forces, are needed to address the challenges of warfare in the ensuing decades.

---

28 Gordon and Trainor, *Cobra II*, 109-117.
29 Gordon and Trainor, *Cobra II*, 586-585.
30 Gordon and Trainor, *Cobra II*, 546-547.

Moreover, expeditionary warfare fundamentally means offensive warfare; for why would a force deploy across the world to immediately assume a protective posture? If the assumption that expeditionary warfare is synonymous with offensive warfare, then the 3-to-1 (3:1) attack-to-defense ratio should remain at the fore for how the US military envisions future force design.[31] Certainly, a small degree of that ratio can be overcome through technological innovation, but history suggests that at the end of the day, success in land warfare resides in numbers, and not technology.

The light, small force looks attractive for bean counters, those in charge of the purse strings, and the individuals attempting to gain influence over the bean counters and those who hold the purse. History, however, shows that a small, light force might accomplish an immediate war aim, such as toppling a weak government, but they do not provide military commanders flexibility to address the military and civil challenges that quickly emerge after initial, albeit indecisive military success. Moreover, small, light forces, unable to address the litany of land warfare challenges that emerge in the subsequent phases of expeditionary operations, will result in the deployment of more land forces, or, the outsourcing, or contracting, of land force operations to third party proxies, to overcome personnel shortcomings. Therefore, if the small, light force Paradox wins the debate, the US military, and its Western partners, must be prepared to manage an increasing number of proxy wars moving forward. Lastly, the challenge-response cycle suggests that any initial benefit associated with small, light, and more deployable land forces would quickly be matched, and overcome by thought adversaries interested in winning, and not supporting ill-advised Western military force design.

Paradox 3: The Looming Importance of Battlefield Transparency

Analysts and practitioners assert that the transparent battlefield is both an emerging concept in warfare and that it will revolutionize warfare. The belief is that sensing capability is becoming increasingly relevant on battlefields because of the proliferation of drones, space-based monitoring capabilities, and small land-based sensors.[32] As a result, military forces will be under constant observation and, therefore, prone to enhanced and timely

---
31 Trevor Dupuy, *Understanding War: History and Theory of Combat* (London: Leo Cooper, 1987), 174-175.
32 *Frontline Podcast: The Transparent Battlefield*, 28 November 2022, accessed 1 September 2023, available at: https://www.youtube.com/watch?v=1AFEeok9vh0.

targeting and strike.[33] Analysts David Barno and Nora Bensahel go so far to emphatically state that, "The future transparency of this expansive web of support should be nothing short of terrifying to U.S. military planners… These factors have staggering implications for future Army doctrine, organizations, and platforms."[34]

Yet, any generation that experiences significant technological or process advancements in reconnaissance, surveillance, sensing, and improvements in strike capability and process, tends to find commenters quick to assert that surprise is a vestige of a bygone era of armed conflict. This is because, as commenters suggest, technological innovation has created a transparent battlefield in which military commanders and the intelligence community possess nearly infinite information on the operating environment and the forces located therein. Analyst Wilf Owen, for instance, correctly highlights that the transparent battlefield has been a feature of armed conflict since at least World War I, and battlefield transparency is an evolutionary aspect of warfare.[35] Brigadier General Curtis Taylor also notes the historical precedent of battlefield transparency by referring to the use of aerial observation balloons and other information gathering capabilities dating to the nineteenth century.[36] Writing in 1993, B.R. Isbell records that theorists from as far back in time as the nineteenth century, such as the indominable Antoin Jomini, have written that the growing number of battlefield sensors has rendered surprise all but obsolete on transparent battlefields.[37]

First, it is important to remember that maintaining a higher number of a given type of system equates to dominance in that field. Take military satellites, for instance. According to the website *World Population Review*, the top three states for military satellites are the US (239), China (140), and

---

33 Andrew Eversden, "Wormuth: Here are the 6 Areas the Army Must Prepare for in 2030," *Breaking Defense*, 15 September 2022, accessed 1 September 2023, available at: https://breakingdefense.com/2022/09/wormuth-here-are-the-6-areas-the-army-must-be-prepared-for-in-2030/.
34 David Barno and Nora Benasahel, "The Other Big Lessons That the U.S. Army Should Learn from Ukraine," *War on the Rocks*, 27 June 2022, accessed 1 September 2023, available at: https://warontherocks.com/2022/06/the-other-big-lessons-that-the-u-s-army-should-learn-from-ukraine/.
35 William Owen, 'The False Lessons of Modern War: Why Ignorance is Not Insight,' *The British Army Review*, Vol. 183 (2023): 26. Available at: https://issuu.com/chacr_camberley/docs/british_army_review_183?fr=sNDM3MzUzOTM1NTQ.
36 *Frontline Podcast: The Transparent Battlefield*, 28 November 2022, accessed 1 September 2023, available at: https://www.youtube.com/watch?v=1AFEeok9vh0.
37 B.R. Isbell, "The Future of Surprise on the Transparent Battlefield," in Brian Holden Reid ed., *The Science of War: Back to First Principles* (London: Taylor and Francis Group, 1993), 147.

Russia (105).[38] No other state comes close to Russia's third place position. The high level of classification of space-based military capability, however, makes it challenging, if not impossible, to validate these numbers. Nonetheless, *World Population Review's* numbers are used within this section to help illustrate ideas about the transparent battlefield.

Second, dominance, from purely a quantifiable standpoint, does not always translate to applied dominance. To be sure, history is replete with examples in which more of something, whether that be larger forces, more artillery, or more drones in the sky, does not pave the way to battlefield, or political, victory. A similar understanding should be applied regarding the future of sensing and its impact on the conduct of warfare. Similar to the points raised regarding Paradox 2, the challenge-response cycle suggests that as a state develops more sophisticated means by which to observe the battlefield, and the likely routes expeditious forces will take to those battlefields, the adversarial state will make equal investments in ways to confound sensing technology.

Third, Robert Leonhard cautions that limited technologies, or those that significantly extend the range for which a military commander can generate a high-impact battlefield effect, are often retained at higher levels of command. In non-Western militaries, in which freedom of action is not necessarily encouraged, nor rewarded, one must assume this situation is even more prevalent. The assumption must then be made that a battalion or brigade, regardless of its status as Western or non-Western, is not going to be privy to sensing capabilities, nor necessarily sensing information that does not directly, or peripherally, impact its operation. A theater commander or field army's sensor feed would no doubt overwhelm a battalion hundreds of miles forward, especially if it was doing anything above or beyond passively occupying an uncontested position on the battlefield.

Resultantly, battlefield transparency will provide momentary bursts of military advantage, but once the technology associated with those breakthroughs are identified, the respective adversary will find additional methods through which to again obviate transparency. Therefore, Western militaries should continue to invest in deception techniques and technology, as well as counter sensor techniques and technologies. Yet, at the same time,

---

38 "Military Satellites by Country 2023," *World Population Review*, accessed 1 September 2023, available at: https://worldpopulationreview.com/country-rankings/military-satellite-by-country.

Western militaries should also see battlefield transparency as just another means of 'combined arms' and 'joint' being applied to the battlefield.

However, the transparent battlefield does warrant a degree of caution, especially for maneuver enthusiasts. Maneuver warfare on a transparent battlefield might well be suicidal. Taking the logic of transparency beyond the emotional stage of recognition finds that positional and attritional warfare are likely the inevitable byproducts of unchecked technological innovation. For instance, if a transparent battlefield becomes a reality, a force's (Force A) dashing movements across unrestrictive terrain, in which an adversary (Force B) possesses the power to identify, target, and engage Force A in a manner of minutes, or perhaps even seconds, holds the promise of eliminating Force A from participating in a conflict before the conflict even gets underway. Advancing on a transparent battlefield against an adversary equipped with an effective sensor-to-shooter network, would certainly welcome a significant bombardment.

Left unaccounted for, military operations on a transparent battlefield will likely evolve from maneuver-type operating, toward something more akin to positional warfare with a strong attritional flavor. Instead of seeking battle against an enemy force in open terrain, an aggressor might hurriedly move from one location to the next, attempting to find obscurant terrain along the way, while seeking a beneficial position proximal to their adversary – perhaps in a city – to bring their adversary to battle. Transparency, if it does materialize, will turn battle into a positional affair, even more than its present state.

A military force's sensor-to-shooter network might be so prohibitive to movement that both sides move into urban areas and conduct duals of long-range precision strike with one another. In this situation, a force might only be willing to move if their adversary's sense, strike, and protection systems have been eliminated to a condition sufficient for the adversary to use harassing strikes. This situation – movement to destroy with fires – is not a high-minded take on maneuver warfare in a sensor-to-shooter rich operating environment but, rather, the long overdue acknowledgement of positional and attritional warfare's relevance in armed conflict.

Paradox 4: Warfighting Preference Matters

The perceived importance of warfighting preference is a theme that permeates military thought. To be sure, the cult of maneuver is quite strong in Western militaries, despite the concept not delivering on its associated hype. The same holds true for conflicts characterized as irregular wars in which Western militaries are working through 'partners' but, in reality, are proxy wars in which Western militaries are leveraging third-party actors as intermediaries for their own self-interest.

Western militaries tend to be surprised when a conflict, such as the ongoing Russo-Ukrainian War, quickly takes on an attritional hue, or when a proxy conflict emerges from a military operation intended to support a friendly state or non-state actor. Despite a state's best wishes, warfighting tends toward attrition because a) an adversary will always operate in a way that provides it the best opportunity to survive and win, and b) an adversary will operate in ways that suboptimize how their opponent wants to operate and how they are built to operate. Similarly, operating through an intermediary actor, whether that actor is a state or non-state actor, and having them engaging in combat in lieu of one's own forces, in pursuit of one's self-interest is by definition a proxy war. But by obfuscating the reality of proxy actor engagement, Western militaries put themselves at a disadvantage by not being prepared to understand how to appropriately account for agency costs and 'partners' not operating in how the principal prefers.

Reality, on the other hand, demands that a state possesses a multifaceted capability to fight in a host of hostile environments, against, and with, a variety of state and non-state actors. Reality, therefore, demands that a military not vector its understanding of warfare, nor optimize its training, for a preferential way of warfare. A military's insistence that it will not participate in operations in urban environments, proxy wars, or that they will not engage in attrition warfare, for instance, are a few common preferential proclamations offered by Western militaries.

Nonetheless, these proclamations are hollow aspirations. Five factors determine how a military must fight: 1) the physical environment carries a significant amount of determinism. Many physical environments, such as urban areas, heavily wooded areas, or waterways, inhibit mobility and thus deny maneuver warfare beyond the smallest tactical elements. Try as one may, if the physical environment dictates a plodding methodical way of operating, then maneuver must remain an aspiration. 2) time, although not as deterministic as terrain, puts pressure on how a military force will operate. Time is inextricably linked with opportunity, and occasionally time rewards the military most responsive to potential opportunity, and not perfected planned and resourced operations. 3) a military force must consider its adversary. An adversary's method of fighting, the location in which they elect to fight, and their respective force design all influence how a military must account for that adversary on the battlefield. 4) and finally, one's preference for fighting can be considered. Moreover, one's preference should not be considered ahead of any of the four variables, but instead, in coordination with the other factors. Not taking the other considerations – terrain, time, an adversary – seriously, and assuming that fiat can overcome reality, is a fool's errand.

Paradox 5: Defeat Mechanisms

US Army doctrine defines Defeat Mechanisms as, "The method through which friendly forces accomplish their mission against enemy opposition."[39] Defeat Mechanisms are communicated through the application of one of four tactical mission tasks, to include: destroy, dislocate, disintegrate, and isolate.[40] Two significant problems, however, make the Defeat Mechanisms incompatible with the trajectory of armed conflicting moving into the future.

First, the Defeat Mechanism is a mechanical heuristic, developed for a period in which warfare was less dependent on the interconnection between individual implements of warfare and more dependent on the communication between echelons of command. Moreover, Defeat Mechanisms were not borne from analytical rigor, but instead they were crafted by doctrinaires at Fort Leavenworth as a cheap and easy tool to help planners, staffs, and commanders communicate how they envisioned tactically defeating an adversary.[41] Frank Hoffman and Eado Hecht have attempted to add intellectual rigor to Defeat Mechanisms, but that work is *post facto*.[42] In future warfare, however, artificial intelligence, machine learning, autonomous systems, and many other information-driven, network-based tools will force the concept of defeat to exist on a plane greater than the purposeful impact on an adversary's physical tools of warfare. Instead, defeat will result from a state's ability to exhaust its adversary's warfighting system through interconnected operations

---

39 Field Manual 3-0, *Operations* (Washington, DC: Government Printing Office, 2017): 1-21.
40 FM 3-0, *Operations* 1-21 – 1-22.
41 Comments reflect a month-long period of search on the subject that occurred during October 2022. The research involved email and phone calls with individuals at Fort Leavenworth's doctrine directorate and Fort Moore's doctrine directorate.
42 See Frank Hoffman, "Defeat Mechanisms in Modern Warfare," *Parameters* 51, no. 4 (2021): 60-64; Eado Hecht, "Defeat Mechanisms: The Rationale Behind the Strategy," *Military Strategy Magazine*, Vol. 4, no. 2 (2014): 24-30.

targeting the adversary's physical force, the information it needs to make decisions, and the time it requires to make effective decisions. In this systems warfare environment, in which system exhaustion is the surest path the battlefield success, Defeat Mechanisms are passé.

Second, considering the importance of systems thinking, Defeat Mechanisms fail to reconcile how networks and information-driven systems insulate battle command – from the tactical to the strategic levels – from physical destruction. Destroying a computer on a network, or destroying a tactical command post, for example, does not destroy the information on the network. Destroying the computer, or the command post, only delays information from reaching its intended user on that network. As a result, the data on the network, not the physical manifestations of the network (for example, computers, servers, command posts, and so on), and the network's feedback loops, should be the focal point for Western militaries seeking to defeat adversaries on future battlefields.

Considering the importance of systems thinking on how states and non-state actors operate in armed conflict, it is imperative to understand that networks are defeated when data on the network is not trusted. Period. Of secondary consideration, networks are defeated when the information needed to make the network work moves in insufficient quantity through the network to fuel the feedback loop process.

Moreover, networks are defeated when the information on the network flows at too great a speed for the network, or the humans on the network, to manage the flow of information. Lastly, networks are defeated when the information on the network is corrupted to the degree that it presents a false sense of reality, and therefore, encourages poor decision-making in an adversary.[43] The Defeat Mechanism heuristic, which is anchored on attacking an actor's physical means, and does not reflect the realities of system warfare, is a Paradox much like the transparent battlefield, which comes with vampiric qualities in that it seems to resurface whenever Western militaries begin examining how to operate in armed conflict.

Defeat Mechanisms, the product of sequential, mechanistic thinking about armed conflict, do not appropriately account for time in armed conflict. Time's importance to systems thinking and systems warfare cannot be overstated. Andrew Carr correctly posits that savvy commanders use fast or slow tempo operations to bend time to their advantage and help

---

43 Donella Meadows, *Thinking in Systems: A Primer* (White River Junction, VT: Chelsea Green Publishing, 2008), 39-50.

accelerate an adversary's defeat.[44] Considering the importance of systems thinking in the future of armed conflict, time holds a germane position as it relates to military and strategic defeat. Time is manipulated in warfare – systems warfare or otherwise – through the tempo of operations. Speed, or operating with high tempo, is not always the best operational or tactical solution to a military problem.[45]

With defeat as the watchword, the Defeat Mechanisms construct should be cashiered for a heuristic better suited for the realities of modern and future armed conflict. As a result, any resulting warfighting heuristic should suboptimize around the ability to fight joint, apply combined arms, and operate in multiple domains to the point at which defeat is obtained. Accordingly, defeat is generated by exhausting the enemy's warfighting system, not through the destruction, isolation, dislocation, or disaggregation of any part (or parts) of their fielded forces. Defeat is generated through the combined effort of three pillars of activity – physical action, data and information manipulation, and time manipulation. Each pillar consists of its own set of basic tasks, and they are linked through interwoven tracts that advance to defeat. As the use of the phrase *interwoven tract* implies that the progression to defeat is not linear, nor binary, but a responsive heuristic that is sensitive to changes in the environment. However, to generate defeat on future battlefields, a combatant must mutualistically use the interwoven tracts of physical destruction, data and information manipulation, and tempo exploitation to exhaust an adversary, and thus bring about an adversary's operational or tactical defeat.

Figure 3.1 provides a graphical representation of the interwoven tract alternative to Defeat Mechanism. In Figure 3.1, the X-axis represents a notional phasing construction to illustrate how physical, data, and temporal elements of warfare can be synchronized to create a general phasing construct. Initiation, Major Combat Operations, Exhaustion, and Defeat are the four major components articulated along the X-axis. These components represent the major benchmarks a combatant might encounter using the interwoven tract heuristic to visualize how to defeat an adversary. Along the Y-axis are general measurements – high, medium, and low. The graph does not quantify any of these measures because that degree of fidelity is not needed to help illustrate the point being made by Figure 3.1. Within

---

[44] Andrew Carr, "It's About Time: Strategy and Temporal Phenomena," *Journal of Strategic Studies* 44, no. 3 (2021): 306-308.
[45] Olivier Schmitt, "Wartime Paradigms and the Future of Western Military Power," *International Affairs* 0, no. 0 (2020), 2-4.

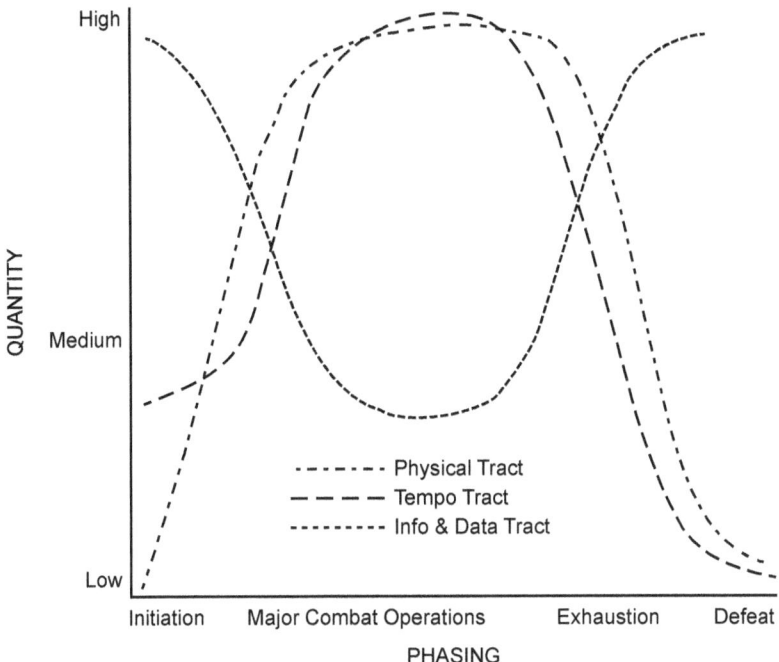

Figure 3.1 Tracts and Phasing over Time

the graph, the physical tract, information and data tract, and tempo tract are illustrated by different types of lines. Their differing position along the X-axis illustrates how a state or non-state actor might utilize each tract, in relation to the other, during each phase of the operation to advance their opponent from the initial conditions of conflict, through major combat operations, and into exhaustion, and subsequent military defeat.

Conclusion

To address the challenges of the future of armed conflict, Western military thought must expand beyond the confines of engrained institutional thinking. It must periodically question its assumptions and its extant mental models. To keep pace with change, Western military thought must discharge obsolete ideas and concepts, regardless of how uncomfortable doing so might feel. Western military thought must move beyond appeals to authority to legitimize its guiding ideas.

Moreover, Western military thought must not fall victim to emotionally reacting to flashy videos on social media to make claims about fundamental changes in war and warfare. Military thought is far too a serious business to allow emotionally charged reactions drive adaptations in doctrine and force structure. As a result, military theorists must periodically review the ideas permeating international relations scholarship, think tank publications, and other forms of media pertinent to the study of armed conflict. They should review these things and ruthlessly seek to eliminate misleading or incorrect information and, in its place, position empirically informed evidence that better represents the truth in both war and warfare.

4 The Principles and Inverse Principles of War and Warfare

Having defined a handful of ideas that are undercutting clear thinking about contemporary and future armed conflict, it is therefore important to re-examine the principles of war. The principles of war should not be a standalone set of tropes, but instead, they must exist to describe the essence of conflict and combat for the time in which they are used. Principles of war, therefore, should be updated periodically to ensure that they are in fact maintaining relevance. The joy about the principles of war is that they are a worldly concept, meaning that no one individual, nor institution, possesses proprietary agency over them and, thus, they can be updated by anyone with the inclination to do so. Moreover, considering that today's principles of war have remained largely unchanged for the past 80 years, it is high time to address the idea.

One might suggest that the US military updated the principles of war several years ago when they introduced their Principles of Joint Operations.[1] The US military did this to account for challenges identified during their post-9/11 wars in Afghanistan and Iraq. However, the US military's emphasis on joint operations, and the inclusion of counterinsurgency-specific language, demoted the principles from a broad general idea pertaining to the conduct of war and warfare, and instead made them applicable only to one type of military operations.

The principles of war generally used by Western militaries today derive from J.F.C. Fuller's work on the subject during the early part of the twentieth century. Fuller toiled over the principles, reworking them at least four times, before settling on what are today's accepted principles of war.[2]

---
1 Joint Publication 3-0, *Joint Campaigns and Operations* (Washington, DC: Government Printing Office, 2023), I-1.
2 See J.F.C. Fuller's The Foundations of the Science of War; "Gold Medal (Military) Prize Essay for 1919," *RUSI Journal* Vol. 65, no. 458 (1920); *Training for Soldiers* (London: Hugh Rees, 1914).

Principles of war, first developed by J.F.C. have remained relatively constant for a century. In recent years, authors have broached the topic, especially as they relate to the future of armed conflict, to varying degrees of success. Despite these theorists attempts to help improve the principles that underpin much of Western military thought and the associated strategic processes, many Western militaries continue the worrying trend of relying on stagnant, centuries old ideas, and easily falsifiable principles and concepts as the bedrock of their thinking.

Nonetheless, principles of war should be based on rationale sourced from the marketplace of ideas and anchored on the attitude that the science of war exceeds the bounds of ideology. That is to say, institutional bias has no place in the development of basic concepts of war and warfare. In that spirit, this chapter posits that five basic principles guide war, and that each of those principles has an inverse principle, in effect describing what a military force must do in armed conflict, and what they must prevent from happening to them. Accounting for inverse principles is important because this technique ensures that self-interest does not cloud the necessity of seeing war and warfare through the lens of multilateral competition in which all participants pursue strategic victory. As theorist B.H. Liddell Hart cautions:

> For war is a two-party affair. Thus, to be practical, any theory must take account of the opposing side's power to upset your plan. The best guarantee against their interference is to be ready to adapt your plan to circumstances, and to have ready a variant that may fit the new circumstances.[3]

The five principles of war are: (1) survive, (2) win, (3) order, (4) durability, and (5) power. The inverse principles are, respectively: (1) extinguish, (2) lose, (3) disorder, (4) embrittle, and (5) starve.

I also introduce principles of warfare to account for what my research indicates are the most salient features in armed conflict that a military force must address. Like the principles of war, the principles of warfare also have inverse principles that must be accounted for.

The principles of warfare are: (1) pragmatism, (2) unpredictability, (3) movement, (4) redundancy, and (5) information. The inverse principles

---
3 B.H. Liddell Hart, *The Ghost of Napoleon* (London: Faber and Faber Limited, 1934), 114-115.

of warfare are (1) idealism, (2) predictability, (3) immobility, (4) essential, and (5) ignorance.

Principles of War

Having previously written on the principles of war, further reflection cautions that a separation between principles of war and principles of warfare is required. Separation is required because of the ontological differences between war and warfare. Christopher Tuck states that war is a state's acts of policy and military strategy that pertain to a specific conflict. Whereas warfare, on the other hand, is the action taken by a state's military force (or a non-state actor) pursuant to the state's war aims.[4] Put another way, war is the strategic considerations of armed conflict, whereas warfare is the operational and tactical considerations, or where actual combat occurs. That structural difference means that causality, importance, and requirements are not universal between war and warfare, and therefore each level requires its own set of principles. '

What's more, inversion applies to the principles of war. Inversion is the result of having changed a principle's direction, or orientation. In terms of photography, think of inversion as a picture's negative. Transitioning this idea to the principles of war and warfare, then, if winning is a principle, for instance, then defeating the opponent is the inverse principle. In practical terms, each principle of war reflects an important feature of armed conflict that a state or their military must accomplish. The inverse principle, on the other hand, reflects what a state must prevent their adversary from accomplishing Therefore, each principle listed within this chapter, resultantly, carries with it an inverse principle (see Table 4.1).

Principle #1: Win

The dialogue on victory or failure in armed conflict has become too esoteric, and likewise, strategic theory, in some cases, has gone too theoretical. In many circles, discussions on victory in armed conflict are bypassed altogether as deliberators trip over academic phrases such as theory of victory, and what that phrase means, more so than generally defining victory.[5] Moreover,

---

[4] Tuck, *Understanding Land Warfare*, 2-3.
[5] Brad Roberts, "On Theories of Victory, Red and Blue," *Lawrence Livermore National Laboratory*, Livermore Papers on Global Security No. 7 (2020): 26-39.

Principle	Inverse Principle
Survive	Extinguish
Win	Lose
Order	Disorder
Durability	Embrittle
Power	Starve

Table 4.1 Principles and Inverse Principles of War

strategic theory is on the verge of jumping the shark as theorists push the bounds of logic by refusing a resource-bounded reality, instead offering that victory is armed conflict, is of no concern to the strategist because the acme of strategy is not the finality often linked with victory, but rather, remaining engaged in the strategic competitive environment.[6]

Of secondary importance, is to understand that strategic actors define their own definition of victory, and definitions of victory are gradient scales. The Russo-Ukrainian War is instructive for gaining an appreciation of definitions and gradient scales of victory in armed conflict. From the beginning of the Russia's February 2022 re-invasion of Ukraine, many so-called experts have jockeyed for prime position in the public eye by making bold predictions about the status of Russia's war in Ukraine. Since the week of the conflict, prognosticators have forecasted that Russia's military was at the precipice of defeat and that Russian society could not, and would not, accept the war's economic impact, nor its need for soldiers to replace losses on the front. Yet most experts have been wrong. Russia's society has weathered the hardships, and its military has moved on from operating the way that it wanted to fight – which did not work and put it in harm's way – and adopted forms of warfare that better aligned with the situation and Russian war aims.

War aims are important to consider when addressing the 'compete to win' principle of war. A state's policy goals for a war – or war aims – are the true benchmark of victory in war. A state's war aims might be in the public record, or they might be kept behind a veneer of truth. Furthermore, war

---

[6] Everett Dolman, *Pure Strategy: Power and Principle in the Space and Information Age* (London: Frank Cass, 2005), 102-103.

aims change as the situation within a war changes. These changes can include heroic victories, horrific losses, the addition or subtraction of partners, the influx or depletion of resources, changes in the attitude of the international community, or any other number of considerations. Nonetheless, one must appreciate that war aims are never static. As a result, a state's war aims align to a gradient scale along which maximum and minimum acceptable outcome are the bookends. Moreover, this gradient scale is likely a feature of statecraft and military affairs that is generally kept from view from the public for military and diplomatic necessity.

This is important to understand because it then makes 'compete to win' a principle that exists on a sliding scale, and so while a state (State A) might think that it is effectively advancing toward victory, it is eliminating the possibility of its adversary (State B) to unlock maximum acceptable war aims. Yet, the effort State A has exerted by this point has reached an inflection point and exhausted State A's ability to aggressively continue the conflict. Concurrently, State B marks its maximalist gains from its play sheet and moves on to lesser war aims, but not less critical to their definition of strategic victory.

Returning to the Russo-Ukrainian War helps illustrate this idea. Synthesizing the amount of forces Russian put into the field against Ukraine and Russia's initial military operations against Ukraine suggests that Russia's war aims in Ukraine were never truly focused on subjugating all of Ukraine, nor militarily defeating Kyiv's military. Instead, Russia's war aims appeared focused on eliminating the sitting Ukrainian government, replacing it with a Kremlin-friendly regime, occupying Ukraine from the Dnieper River, east to the Russian border, and eliminating the remainder of Kyiv's forces that stood in the way of accomplishing those goals. In effect, Russia's war aims were not the annexation of all of Ukraine, but the annexation of Ukraine from the Dnieper River to the Russian border. It is useful to recognize this goal as Russia's general military strategy for its war against Ukraine.

Within Russia's general strategy, a set of sub-strategies broadened and deepened the Kremlin's definition of victory in Ukraine. The retention of the Donbas and Crimea remained central points of emphasis, and thus supporting sub-strategies. The linking of the Donbas and Crimea along the much sought after 'land bridge' was another of the Kremlin's significant strategic aims, and therefore, another of the Kremlin's sub-strategy.

Figure 4.1 Minimum Goals, Ukraine
Figure courtesy of the Association of the United States Army

The interplay of sub-strategies within the context of a general military strategy is important to appreciate because gradients of strategic victory exist in a state, or non-state actor's, reliance on sub-strategies. Stated more clearly, sub-strategies can illustrate a state's decreasing definitions of strategic victory. For instance, if the Kremlin's overall definition of strategic victory was the annexation of Ukraine, from the Dnieper River to the western Russian border, which could be viewed as its maximalist definition of victory, while gaining and maintaining the land bridge to Crimea – and the retention of the Donbas and Crimea – could be viewed as the Kremlin's minimal accepted outcome in the conflict. Kyiv's ability to inflict territorial loss greater than the latter point might be perceived within the Kremlin as strategic failure. Figure 4.1 provides a graphical representation of hypothetical Russia definitions of victory in Ukraine.

Appreciating a state's definition of victory as a gradient scale helps understand that victory in war is not an all or nothing, but instead, it often operates on a scale of acceptability. Not appreciating a state's definition of victory can result in an adversarial state misjudging how close, or how far, they are from victory. But, nonetheless, all states should anticipate their adversaries to fight vigorously for victory, whether that's maximalist general strategy goals, or minimalist sub-strategy aims.

Simultaneously, war – in which death and destruction are the currency in which victory must be purchased – is inherently uneconomical. Indulging in war with no intention of victory is wasteful and a luxury few states can afford. States and militaries not intent on victory squander their resources from across not only the military, but also the diplomatic, information, and economic elements of national power. Therefore, it is imperative for a military force to win, while simultaneously, quickly ushering their adversary across the threshold of defeat.

Principle #2: Survive

Modern militaries are self-aware organizations. Self-aware entities pursue survival above all other goals. In armed conflict, to survive means to continue existing despite environmental challenges, threats, or dangers. Being self-aware and focused on survival in war causes modern militaries to operate in a self-organizing manner, despite their hierarchical structure.

Military organizations place the ability to survive, and subsequently, the requisite self-care, ahead of all other considerations. A military force's access to data and information, controlling the rate of operations, and maximizing its protection, are just as germane to its survival as the force's need to maintain the power to gain proximal, situational dominance at important places and times on the battlefield.

Military forces operating with self-awareness and self-organization indicates the presence of the logic of systems theory in modern military forces, their operations, how they interact with their superior and inferior partners, and how they interact with a dynamic world around them. Feedback loops are the critical link between military forces and all the environmental variables that impact their survival. Considering the logic of feedback loops, in which data and information from the outside world is collected, analyzed, synthesized with internal information, and distributed across the organization, a military force's survival is inextricably linked to the possession of accurate information, the safeguarding of internal information, and the redundancy of data and information networks.

Conversely, extinguish is survives inversion principle. The military force that is adhering to the principle of survival is simultaneously working to extinguish their adversary, just as their adversary is doing the same. Data and information are just as important to extinguishment as they are regarding survival. Therefore, a force seeking to extinguish an adversary should ruthlessly strike at the latter's data and information and their information processing systems.

Principle #3: Order

The concept of entropy, in which all systems tend towards disorder, is a key idea pertinent to military forces and the practice of war and warfare. Military forces – both in garrison and combat – are always moving towards a state of disorder. Regarding militaries, entropy presents itself as a degradation in capability (for example, the physical ability) to accomplish tasks and missions, and in the depreciation of capacity (for example, the people and organizations) to achieve objectives and fulfill assignments. In self-organizing, self-aware, and open, learning systems, such as modern state militaries, entropy is not allowed to run rampant. Militaries seek to offset the constant repercussions of entropy through tasks such as recruiting, training, proactive medical care, preventative maintenance, sustainment activities, and education.

The relationship between entropy and military forces is a cycle, in which military forces seek order, whereas the natural force of entropy fuels disorder (see Figure 4.2). One way to visualize this idea is to place 'entropic effect' and 'countering entropic dynamics' as poles in a cycle chart, and labeling former 'Node 1' and the latter 'Node 2'. The arm that connects Node 1 to Node 2 are all the organically occurring activities inherent to entropy in military forces, plus in war and warfare, all the adversarial injected activities to further accelerate disorder in an opponent. The arm that connects Node 2 to Node 1 consists of all the naturally occurring tasks that a military force does to overcome naturally generating disorder, plus everything that it does to compensate for and overcome adversarial caused disorder.

As Figure 4.2 alludes to, disorder is a military force's natural state of existence. The devolution to disorder accelerates when an adversary is added to the equation. Thus, war is the process of military force to overcome the natural and adversarial generated components of disorder through attentive monitoring of environmental and internal conditions, to modulate their own activities to counter entropy's negative impact on their ability to accomplish tasks and complete missions. Therefore, if disorder is a military force's natural state of being, then order is a military force's foremost objective and the pursuit of order its main concern.

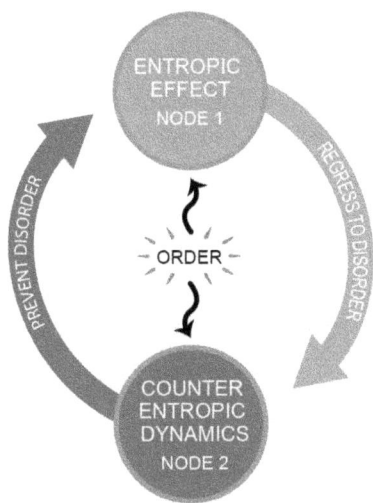

Figure 4.2 Entropy Cycle
Figure courtesy of the Association of the United States Army

The polarity of order and disorder thus provides the principle of order and inverse principle of disorder in war. In follows that a military force must always operate in ways that advance their own order and accelerate disorder in their adversaries. Combat operations are one way to intervene in the order-disorder cycle, but battlefield architecture and force design are additional ways to influence order and disorder in war.

Principle #4: Durability

Durability, in the context of armed conflict, is a military force's capacity to absorb shocks, recover quickly from adverse situations or conditions, and maintain its structural integrity. Beyond the physical perspective, durability figures into applied situations. That is, military forces must be able to operate in austere environments for extended periods of time in order to win their respective operation and survive with sufficient power to maintain order over itself and the emergent situation.

Durability is a principle of war, especially for Western military forces, because they are generally required to conduct expeditionary military operations. Considering this factor at the outset of military operations, Western military forces often conduct offensive operations against fortified adversaries determined to protect their forces and their interests from invading Western military forces. Given that adversary forces are likely defending important terrain, such as seaborne-landing sites, airfields, and potential drop sites, one can then assume that materiel destruction and loss of individual soldiers in those critical locations will be increasingly high. Therefore, Western military forces require durability to sustain contested expeditionary operations but, equally important, Western military forces require durability to conduct objective-oriented operations after successfully landing at embarkation points.

The principle of durability also obviates the idea that smaller, lighter forces are the answer to contemporary and future challenges in armed conflict. The principle of durability suggests that military forces – especially those on the sharp edge of tactical warfighting – should not be designed with slim margins, but rather, they should be designed and outfitted with sufficient personnel, equipment, warfighting systems, and units to overcome the rigors of war against an adversary determined to win.

Further, the urbanization of warfare is a prevalent trend in modern war. Noted Anthony King also provides context, writing:

> Force size may seem banal and unimportant, but in fact it has typically played a significant role in the character of military operations…In the twentieth century, when states possessed massive armies, they were able to dominate urban areas through force of numbers. However, as forces have downsized, states have struggled to control urban areas.[7]

Moreover, King contends cities offer adversaries, and weaker opponents, "The best opportunities for evasion, concealment, ambush, and counterattack against the technologically superior weaponry of state forces."[8] King's arguments, fresh among scholars and practitioners, suggests that smaller, lighter forces are not the answer to problems of current (and future) war, but rather larger and more durable forces are required.

---

[7] Anthony King, "Urban Insurgency in the Twenty-First Century: Smaller Militaries and Increased Conflict in Cities," *International Affairs*, Vol. 98, no. 2 (2022): 609-610. Doi: 10.1093/ia/iiac007.

[8] Anthony King, "Will Inter-State War Take Place in Cities?," *Journal of Strategic Studies*, Vol. 45, no. 1 (2022): 69. Doi: 10.1080/01402390.2021.1991797.

Embrittlement is durability's inverse principle. If maintaining durability is required to address the significant challenges of both expeditionary and joint force supported land warfare, then avoiding brittleness is important. Conversely, deceiving an adversary into situational embrittlement is a key principle of war as well. In the applied sense, embrittlement is arriving for a conflict ill-suited for the environment, threat, temporal, and situational character of a specific conflict. If properly anticipated and acted upon, an embrittled adversary can quickly move through their available resources or be in a physical position in which they are unable to access resources or other means of support, and thus be more prone to military defeat.

Principle #5: Power

Power, in the context of principles of war, relates to possessing the bases of power required to enable campaigns, operations, and battles. Power is the physical means and supporting systems that enable and sustain military operations. Depth in power allows a military force to exhaust an adversary. Depth in power is directly proportional to a force's existing bases of power, coupled with its state's emerging, and latent, bases of power. Power and bases of power have proven their strategic importance in the ongoing Russo-Ukrainian War. At 2024, Ukraine's ability to energize emerging bases of power throughout the international community, as well as Kyiv's ability to mobilize latent sources of power through a variety of domestic reforms, has allowed the state to fight Russia to a draw.

Russia, on the other hand, initially relied on its own internal base of power. However, as Kyiv's military forces thwarted the Kremlin's attempted putsch in Kyiv and pushed Moscow's forces into isolated positions in the Donbas and southern Ukraine, Russia too tapped into emerging and latent bases of power to energize its forces and maintain the ability to keep a grip on its territorial acquisitions in eastern and southern Ukraine. Russia had to reach out to China, Iran, North Korea, and other smaller states for military support. Nonetheless, as the Russo-Ukrainian War illustrates, power is a critical principle of war and one that enables a state's ability to engage in armed conflict.

Starvation is power's inverse principle. Starvation is a state's ability to choke off an adversary's access to the richness of its base, or bases, of power. By starving an adversary's access to its bases of power, a military

> **Box 4.1: How to Apply the Principles of War**
> In war, a combatant must operate with political and military survival (survive) at the fore of their policy and military strategy. Economic considerations, domestic politics, and the uncertainty of time dictate that a combatant must always engage in conflict to win (win), for operating any other way accelerates that combatant toward their own strategic exhaustion. War is fraught with private and incomplete information, making them prone to deception and disorder. Combatants must therefore devise flexible policies and military strategy to account for the impact of imperfect information (order). War operates in an environment of attrition. Combatants must account for attrition in their policy, strategy, associated military activities (durability). Finally, firepower is the variable in war that dictates who wins and who loses, therefore, it is best to possess the most and the best firepower on the battlefield (power). Bases of power – existing, latent, and potential – must exist along a combatant's strategic depth. Those bases of power must be able to quickly mobilize to allow a combatant to generate power sufficient to identify, address, and overcome the challenges of land warfare against powers of equal and greater strength.

force can cause the adversary to become brittle and prone to exhaustion. In turn, those features are certainly not silver bullets, but as history has demonstrated, they do hasten a state's force toward defeat.

Lastly, the principles of war are not a disparate collection of words, but rather the reduction of a basic statement for how states and their military forces should approach war. In thinking about how to apply the principles of war, the narrative in Box 4.1 is a useful tool.

The inverse principles of war are subject to the same guiding framework as the principles of war. Moreover, the inverse principles and the principles of war must be seen as interlocking principles that reinforce the other. A framework for thinking about the inverse principles of war in an applied sense is in Box 4.2.

> **Box 4.2: Framework for Thinking about the Inverse Principles of War**
>
> In war, a combatant must conduct military activities directed at denying their adversary's political and military solvency. Doing so will accelerate their defeat, either through the destruction of their forces and political structure or by threatening it such that their political leaders move to negotiate an end to the conflict (extinguish and lose). Chaos, or at least the absence of order, is the general state of being for military forces, which is one of the major reasons why chains of command, communication processes, and training exist. Military operations must endeavor to deprive an adversary of order, which will increase their incoherence and uneconomic activity, accelerating them towards operational and strategic exhaustion (disorder). Lastly, military operations must strive to disrupt and deny an adversary's ability to activate bases as well as operate as a networked system. This will increase fragility throughout the adversary's networked operating system and deprive it of the resources to withstand the rigors of military campaigns (embrittle and starve). In turn, these actions will accelerate the adversary towards operational and strategic exhaustion, and oneself closer toward military and political victory.

Principles of Warfare

The principles of warfare should be waypoints for how military forces operate when engaged in war. Moreover, the principles of warfare should be easily identifiable in a military force's strategy, concepts, plans, operations, doctrine, and activities. Perhaps more important, the inverse principles of warfare should be easy to identify in one's strategy, concepts, plans, operations, doctrine, and activities. By being easy to identify, and written in plain English, the principles and inverse principles of warfare help guide military forces along the proven path of military and political victory in a conflict (see Table 4.2).

Principle	Inverse Principle
Pragmatism	Idealism
Unpredictability	Predictability
Movement	Immobility
Transitions	Paresis
Information	Ignorance

Table 4.2 Principles and Inverse Principles of War

Principle #1: Movement

Borrowing from Trevor Dupuy's thoughts on mobility, which he states are "The ability of military unit's to move as effective formations from place to place...mobility is meaningful militarily only in terms of the relative movement capabilities of opposing forces," movement is defined as the capability to move at operational, tactical, or micro-tactical distances without external support or augmentation.[9] Movement is the fundamental building block for every aspect of warfare. A force cannot advance without the ability to move. A force cannot conduct a coordinated defense without the ability to move. Likewise, a force cannot conduct maneuver or positional warfare without being able to move.

Movement is a fundamental element of warfare. An immobile military force is ripe for attack and destruction. It is thus paramount for a force to possess inherent movement capability and not be bound by either dependency on another organization for movement, or dependency on another service, such as an air force or naval force, for movement. In practical terms, many light forces that have no organic movement capability are not the most useful forces. To be sure, most Western light infantry brigades have to be transported to battle by a mobile force and then they have to be outfitted with vehicles or remain dependent on others for tactical and operational movement across a battlefield. The same holds true for airborne forces. By not possessing sufficient lift or transport aircraft to support independent operations, their utility is not maximized.

---

9 Trevor Dupuy, *Developing a Methodology to Describe the Relationship of Mobility to Combat Effectiveness* (McLean, VA: Historical Evaluation and Research Organization, 1966), 3.

The ability to move quickly in warfare is also extremely important because it allows a military to either take advantage of a fleeting temporal opportunity or create their own situations of advantage relative to their adversary. This is a point echoed by JFC Fuller, who asserts that "Mobile direct protection is generally the most effective, for any change in location necessitates a change in the enemy's tactical organization, and consequently a loss of time for destructive effect."[10]

Principle #2: Pragmatism

In warfare, adhering to a singular way of thinking about how to address military problems is a very dangerous proposition. Warfare, by its nature, is chaotic and ever-changing. Further, warfare is subject to the rules of reality, which manifests in varying degrees of determinism. For instance, if a theater of war is populated by mountainous terrain intermixed with several lakes and rivers – like the US World War II's Italian Campaign or NATO force's disastrous operations during the Korean War's Chosin Campaign – all the forces involved in the conflict are subject to geographical determinism.

Terrain causes a military force to operate most often along the road network. Room for exception applies. If a combatant is composed of non-vehicular forces, they can mitigate the impact of geographical determinism by operating off the road network and moving, albeit at a slow rate, through rough terrain on foot. A force 'whose fighting elements are motorized or mechanized, on the other hand, must operate along the road network because of their vehicle's inability to navigate through rough terrain. In this scenario, the vehicular formation is more powerful, from a capability's comparison, but the terrain all but nullifies those capability advantages. The slower dismounted force, however, is better able to close with the road-bound vehicular force and destroy it with anti-vehicular weapon systems from hidden locations in rough terrain. The same comparison is valid if applied to urban terrain.

This dynamic is why Azerbaijan's military forces appeared indomitable during the 2020 Nagorno-Karabakh War. The Azeris' use of airpower, in this case, armed drones, while the Armenian forces operated road bound mechanized forces and light infantry in mountainous terrain, allowed the Azeris to gain and maintain movement and firepower asymmetry

---

10 JFC Fuller, *The Reformation of War* (New York: E.P. Dutton and Company, 1923), 27.

in the war because they avoided the perils of terrain determinism. As a result, the Azeri military was able to make quick work of the Armenian military on the ground.[11]

Nonetheless, the example illustrates that preference and idealism regarding how to fight falls prey to the deterministic impact of terrain. Other factors such as time, the tactical or operational situation, the forces one has at their disposal, and the enemy's activities within an area of operation, all work together to necessitate pragmatism.

Pragmatism is thus defined as possessing the will, knowledge, and skill to do what a military situation requires, while not being wedded to idealistic or dogmatic prescriptions. Understanding a situation is not just looking for similarities associated with doctrinal templates and then applying an institutional solution. Rather, pragmatism requires unshackling a military force from a prescriptive doctrine and mindset. That force must instead possess a strong appreciation for the variety of warfighting techniques that they might encounter, as well as how terrain, time, and adversarial military activities all contribute to the situation's shape. Beyond just understanding the range of potential battlefield challenges a force might confront, they must also possess the skill, knowledge, and capability to fight within the situation to do what is situationally appropriate to survive and win against an adversary. Dogmatic adherence to idealistic views on war and warfare, which come on the back of axiomatic statements is unhelpful for military forces, both institutionally and in the field. Fuller reminds the student of war "The brain of its commander must in no way be hampered by preconceived or fixed opinions…it is the first duty of the commander to concentrate on common-sense, and to maintain his doctrine in solution so that it may easily take the mould of whatever circumstances it may have to be cast in."[12]

US General Chris Cavoli, Supreme Allied Europe and Commander, US European Command, for instance, is noted for stating that precision beats mass.[13] Cavoli has made this comment in regard to Russia's method

---

11 Phillip Andrews, "Lessons from the Nagorno-Karabakh 20202 Conflict," *Center for Army Lessons Learned* Catalog 21-655 (2021): 3-6.
12 Fuller, *The Reformation of War*, 23.
13 Michael Peck, "Losses in Ukraine Are 'Out of Proportion' to What NATO Has Been Planning For, the Alliances Top General Says," *Business Insider*, 5 February 2023, accessed 28 December 2023, available at: https://www.businessinsider.com/ukraine-war-scale-out-of-proportion-with-nato-planning-cavoli-2023-2.

of warfare 'in the Russo-Ukrainian War. Cavoli implies that with sufficient quantity of precision munitions and precision strike capability, Ukraine could militarily defeat Russia. The problem with this sentiment is that it is unproven and he Russo-Ukrainian War is proving that it is incorrect. To be sure, at the time of writing', despite the use of inordinate quantities of precision munitions, Ukrainian forces are no closer to expelling Russian land forces from the Donbas, the 'Land Bridge to Crimea', nor the Crimean Peninsula.[14] Moreover, Kyiv is no closer to forcing the Kremlin to negotiate an end to the conflict, despite the massive amounts of precision strike capability and precision munitions that the US and other Western states have provided to Ukraine since the start of the conflict. It appears that Cavoli is wrong – mass does overcome precision.

US General James Rainey has made similar idealistic statements. Rainey often states that the US Army "Does not do attrition."[15] According to Rainey, US Army uses maneuver warfare because it cannot, nor is interested in, engage in a one-for-one exchanges in casualties on the battlefield.[16]

Rainey's comments are ironic when one considers his own combat experience as a battalion commander during Operation Iraqi Freedom's Second Battle of Fallujah (7 November – 23 December 2004).[17] Fallujah is the archetype of a battle of attrition – the US military's objective was the elimination of a non-state military force, and the technique to do so was destruction-based warfighting that sought to kill all the fighters, destroy their cohesion, and destroy any buildings within the city that they used as protection and command and control locations.[18] The battle tallied more than 300 coalition casualties, 1,500 enemy combatants killed, 800 civilians killed, and that 60 percent of the city's buildings were damaged, and another

---

14 Angelica Evans, et al., "Russian Offensive Campaign Assessment," *Institute for the Study of War*, 27 December 2023, accessed 28 December 2023, available at: https://www.understandingwar.org/backgrounder/russian-offensive-campaign-assessment-december-27-2023.
15 "AUSA Coffee Series – GEN James Rainey – US Army Futures Command," *Association of the United States Army*, 14 December 2023, accessed 28 December 2023, available at: https://www.youtube.com/watch?v=i_6zFKK3LoA.
16 "AUSA Coffee Series – GEN James Rainey – US Army Futures Command.".
17 "Interview with Lieutenant Colonel James Rainey," *DVIDS*, 16 November 2004, accessed 28 December 2023, available at: https://www.dvidshub.net/video/86271/lt-col-jim-rainey.
18 John Spencer, Liam Collins, and Jayson Geroux, "Case Study #7 – Fallujah II," *Modern War Institute*, 25 July 2023, accessed 28 December 2023, available at: https://mwi.westpoint.edu/urban-warfare-case-study-7-second-battle-of-fallujah/.

20 percent outright destroyed.[19] Fallujah is but one data point in a long line of brutal attritional battles and attritional wars that the US and its partners have fought in in the post-9/11 period. Attrition is a characterization of conflict in which the military objective is the destruction of an adversarial combatant. Idealistic assertions about how to think about, equip for, and train for conflict, like Rainey's regarding attrition and maneuver, or Cavoli's about precision and mass, leave military forces wanting when they encounter situations that do not align with their preferred way of warfare.

Moreover, Cavoli and Rainey's comments are out of step with the true character of war and warfare. Nolan cautions that the historical record illustrates that wars are won by attrition and exhaustion, and that 'Great Captains' or revolutionary methods of warfighting like precision strike strategies occupy a very infinitesimal point in within a deep and broad study of war and warfare.[20] Wars are fought and won through attrition and exhaustion. Therefore, idealistic proclamations about how a force does or does not fight, or that technology can overcome long-standing truths in military thinking, and applied military strategy and operations, is troublesome, if not dangerous. It is dangerous because it can cause states to invest in the wrong technologies, turn off the production of proven warfighting systems, develop improper force design, and incorrectly educate their force for the realities and rigors of armed conflict. Pragmatism must be at the fore of thinking, training, and executing on the battlefield.

Principle #3: Unpredictability

Patterns are one of the easiest ways to think and act ahead of potential adversaries. Operating in a way that creates patterns, whether at a strategic or tactical level, is dangerous because it allows an observant and thoughtful adversary to identify many things – fielded forces, supply nodes and distribution points, command elements, and common routes of supply and advancement. Western military doctrines contribute to the challenge. The US military, for instance, relies unofficially on the phases of joint operations as a simple heuristic for planning and executing military operations. At the tactical level, professional military education in Western militaries often teaches officers an elementary-level sequence of offense and

---

19 Spencer, Collins, and Geroux, "Case Study #7 – Fallujah II."
20 Nolan, "Allure of Battle," 576XXX.

defensive operations. When applied on the battlefield, these tools create problems because they remove a degree of uncertainty that an adversary would otherwise have to address. Statements like those of Cavoli and Rainey contribute to the problem of certainty.

On the other hand, military forces must strive to create uncertainty in their adversary. They do this by operating in unpredictable ways or with unpredictable weapons systems. Operating in unpredictable ways can be achieved by not adhering to things like the phases of joint operations or the sequence of the offense (or defense), or by 'doing maneuver warfare' and relying on precision strike. Further, unpredictability can be achieved by operating according to seemingly odd timings – accelerating the tempo of operations and tactical activity, or on the other side of the token, dragging the pace of operations down to an irregular tempo.

Moreover, applied combined arms theory is paramount in operational and tactical warfighting. However, the mix and application of arms beneath the umbrella of combined arms theory can also be manipulated to create what appears to an adversary as an odd and unpredictable scheme of military activities. What's more, using a combat arm in the place of another combat arm to create the effect of the latter is another example of how manipulating combined arms theory can create unpredictability. This idea can be thought of as the substitution principle of combined arms theory. This idea can become quite heady without a few tangible examples, and therefore it is illustrative to briefly examine Iraqi defenses during the US's 2003 invasion of Iraq, and how Chechens greeted the Russian armed forces in Grozny in 1995.

The Iraqi resistance to the 2003 US invasion of Iraq – both organized government forces and irregular militias – were aware of the perils of using air defense systems to protect against US airstrikes. The Iraqi's understood that if they turned on their air defense systems and engagement US forces and inbound strikes, the US would quickly target those systems.[21] As a result, the Iraqis often resorted to using non-standard combat arms to replace air defense systems, to generate the same combined arms effect of short-range air defense.

A cautionary example of this emerged during the US invasion of Iraq in 2003. The US Army's 11th Attack Helicopter Regiment – the leading

---

21 Michael Gordon and Bernard Trainor, *Cobra II: The Inside Story of the Invasion and Occupation of Iraq* (New York: Vintage Books, 2007), 309.

edge of US V Corps' spearhead fighting north from Kuwait to Baghdad – approached at the twin cities of Haswah and Iskandariyah during the final approach toward Baghdad. Upon arriving on the southern edge of Haswah and Iskandiryah, the regiment found both cities fully illuminated. At 1:00 am, this situation was an odd. Simultaneously, the Iraqis fired high-altitude air defense weapons at the approaching helicopters. However, instead of firing them at high altitudes, the Iraqis fired the air defense missiles at just over 500 feet, or a bit higher than the US helicopters were flying.[22]

To avoid the air defenses', US aviators descended to much lower altitudes.[23] Descending in altitude was exactly what the Iraqis wanted because that brought the US helicopters into striking range of Iraqi small arms fire. The Iraqis defending Haswah and Iskandariyah then unleashed a torrent of small arms and short-range air defense fire on the helicopters. The Iraqi attack quickly overwhelmed the regiment, downed some of their aircraft, and forced the formation to retreat to the safety of the rear area.[24]

The Iraqis' use of signaling, short range air defense, and small arms as a substitute for long range air defense and sophisticated sensor and communications systems is an innovative example of the combined arms theory's substitution principle. Iraqis commanders understood how the US Army wanted to fight – lead with airpower and attack aviation, follow that with cavalry, and then follow through with the main body and support troops. In that regard, US military operations, and specifically the Army's plan, were quite predictable, and therefore a simple challenge for the Iraqis defending that sector of real estate.

The First Chechen War's Battle of Grozny provides another demonstrative example of combined arms theory's substitution principle.

With rebellions rising after the fall of the Soviet Union, the fledgling Russian Federation worked tirelessly to keep its peripheral constituency states intact. Chechnya and other north Caucasus polities, looking to their north and west, observed other nations like Ukraine and the Baltic states exert their right to self-rule and. As a result, Chechnya declared its independence in 1993. The Kremlin quickly mobilized what it believed to be an overwhelming force to address Chechen independence.

---

22 Gordon and Trainor, *Cobra II*, 309.
23 Gordon and Trainor, *Cobra II*, 309-310.
24 Gordon and Trainor, *Cobra II*, 311.

The Kremlin' deployed a large, mechanized land force to capture Grozny and destroy the Chechen's political and military elements within and around the city. Russian military commanders asserted that the entire operation would take 15 days to complete – from start to finish.[25] Grozny's defenders – operating with a clear understanding of combined arms theory – did not use airpower or indirect fire in any serious way against the Russian forces. Instead, they substituted intelligent tactics to compensate for their lack of airpower and indirect fire. The Chechens lured the Russian land forces into the Grozny and then attacked Russia's mechanized forces with anti-armor weapons from the ground and multi-level buildings. Although fighting at a perceived military disadvantage, Chechen operations created the effect on Russian mechanized forces as being attacked from the air and with indirect fire.[26]

Further, the Chechens understood that if they found a way to deny the Russian forces the ability to apply combined arms against them, then they might have a chance of overcoming Russia's military superiority.[27] The Chechens, considering their long subjugation within the Soviet Union, likely understood how Russia would structure their military operations against. In that capacity, the Russian military was a predictable foe whose strengths simply had to be accounted for and offset, making them far from indomitable. As a result, Chechen forces operated in proximity to, or 'hugged', Russian mechanized forces.[28] This technique was effective because Russian forces tended not to use their artillery or airpower for fear of hitting their own forces.[29] What's more, Chechen fighters operated so close to Russian forces that mechanized crews were often not able to operate their tank and infantry fighting vehicle gun because they could not depress the tubes low enough to engage dismounted targets.[30]

Chechen fighters had placed a high toll on Russian participation by the time the battle for Grozny culminated. 'For example, Russia's 131st Motorized Rifle Brigade (MRB) was annihilated. The 131st MRB lost 20 of

---

25 Olga Oliker, *Russia's Chechen Wars 1994-2000, Lessons from Urban Combat* (Monterey, CA: RAND Corporation, 2001), 9-10.
26 Stasys Knezys and Romanas Sedickas, *The War in Chechnya* (College Station, TX: Texas A&M University Press, 1999), 99.
27 Dodge Billingsley and Lester Grau, *Fangs of the Lone Wolf: Chechen Tactics in the Russian-Chechen Wars, 1994-2009* (Fort Leavenworth, KS: Foreign Military Studies Offices, 2012), 171.
28 Billingsley and Grau, *Fangs of the Lone Wolf*, 171.
29 Oliker, *Russia's Chechen Wars 1994-2000*, 19.
30 Oliker, *Russia's Chechen Wars 1994-2000*, 19.

its 26 tanks, 102 of its 120 armored personnel carriers, and all of its anti-aircraft guns.[31] The brigade's commander – Colonel Ivan Savin – and most of his staff were also killed during the battle.[32]

Russia's 131st MRB was not alone. Russia's 506th Motorized Rifle Regiment (MRR), which was one of the primary units supporting the 131st MRB in Grozny, lost more than a quarter of its manpower.[33] The 506th MRR ran into the same innovative Chechen tactics as did the 131st MRB and most other Russian forces in Grozny. By the end of the first month of fighting, Russian combat losses topped 5,000 casualties.[34]

The point of these two lessons in combined arms theory's substitution principle is to highlight the power that unpredictability has on the outcome of engagements, battles, and campaigns. Unpredictable military operations can place a military force in an advantageous position that may well in fact unlock strategic military victory.

Principle #4: Transitions

Transitions are the hinge points in military operations. This is because transitions create phase changes in military operations. The smooth execution of transitions in warfare allows a combatant to maintain constant and exhaustive pressure on an enemy combatant, and thus accelerate them toward exhaustion. Or, the smooth execution of transitions in warfare allows a combatant to disrupt or deny an enemy combatant's constant and exhaustive pressure on themselves, and thus prevent them from culminating because of resource exhaustion.

A simplistic rendering of this idea might be gained by thinking about the transition from offensive to defensive operations. If a transition is appropriately identified and managed ahead of time, then a military force can smoothly move from conducting offensive operations to a defense that accounts for the principles of war and warfare in meaningful ways. If the transition is not properly prepared, inappropriately planned, or perhaps even overlooked, the military force could face ruin as it stumbles through a phase change. Ruin, in this case, is the product of being iteratively churned

---

[31] Knezys and Sedickas, *The War in Chechnya*, 99.
[32] Knezys and Sedickas, *The War in Chechnya*, 101.
[33] Knezys and Sedickas, *The War in Chechnya*, 101.
[34] Arkady Babchenko, "The Savagery of War: A Soldier Looks Back at Chechnya," *Independent*, November 10, 2007, accessed September 8, 2019, https://www.independent.co.uk/news/world/europe/the-savagery-of-war-a-soldier-looks-back-at-chechnya-5329021.html.

through destruction-oriented engagements or battles that attrit manpower, equipment, and other necessities of war and warfare, thereby accelerating the combatant toward culmination by exhausting their resources.

Napoleon Bonaparte provides an important take on the overall importance of transitions to both war and warfare. Bonaparte states: "The secret of war is to march twelve leagues, fight a battle, and march twelve more in pursuit."[35] Though not explicitly stated, Bonaparte's comment attests to the veracity of transitions in the conduct of warfare.

Reading between the lines of Bonaparte's comment finds that the relentless application of destructive and corrosive operations against an enemy combatant compounds the impact of exhaustion. The resulting effect therefore moves an adversary toward culmination quicker than they might otherwise do. Destructive operations are those that destroy an adversary's people and resources. Corrosive operations are those that do not destroy resources, but otherwise generate a suboptimal impact on an adversary. By anticipating, preparing for, and conducting smooth transitions, a combatant can therefore near constant maintain destructive and corrosive operations against an enemy, and thus deprive the enemy combatant of the resources that they require to remain in the conflict, regardless at whichever level it occurs.

The secret of war, according to Bonaparte, is not conducting one of these elements and then stopping. Rather, the secret of war is to anticipate the need to conduct each of these elements and then to conduct them in tandem with one another to maintain constant, exhaustive pressure on an enemy combatant so that they culminate at a time or place advantageous to oneself.

Accepting that hinge points exist in the conduct of campaigns, battles, and engagements is an important first step toward integrating this principle of warfare into one's course of military operations. The hinges – they can be points or phases – are the mechanism through which transitions occur. Further, hinges are born out of situationally dependent conditions. For instance, in Bonaparte's example, the junction between marching twelve leagues and fighting a battle is a hinge point and a transition from one element of warfare to another occurs there. Moreover, the transition of marching to fighting requires a set of conditions to be (a) identified, (b) communicated to that combatants' subordinate elements, and (c) then achieved to be successful. In each element of his statement – movement to

---

35 Alan Schom, *Napoleon Bonaparte: A Life* (New York: HarperCollins Publishers, 1997), 275.

battle, the transition from movement to battle and battle to movement, and the cognitive shift from battle to exploitation – Bonaparte emphasizes the relationship between transitions and the 'secret' in war.

Furthermore, Bonaparte's statement affirms the relationship between momentum through progressive transitions and generating the snowballing effect therein to trigger subsystem and system collapse. Ironically, because of the predictability of operational phasing and sequence at the joint and tactical levels, transitions tend to be known unknowns – an actor is often aware of the required transitions of an operation, but typically does not know when or where they will occur.

Nevertheless, thorough planning can account for much in relation to transitions and reserves, which are two sides of the same coin. The initiation of a transition or the commitment of a reserve must be tied to decision points developed during planning.[36] Finding answers to these decision points must be linked with a system's feedback loop process; it cannot be the sole responsibility of one organization or one capability. '

Paresis is transition's inverse principle. Paresis is theoretically similar to paralysis but differs in that in paralysis an entity does not possess the physical capability to move; whereas with paresis, an entity can move, but it does so at a suboptimal state. The term paresis is used as the transitions inverse principle because realists accept that it is nearly impossible to fully deplete an adversary's ability or physical capacity to move. Or to put it another way, conflict realists understand that creating paralysis, whether physical or mental, is nearly impossible. However, creating a situation in which an enemy combatant cannot move – as has already been stated – is a state of being that one force can impose, through force, on an adversarial combatant. By preventing one combatant from preventing their adversary's ability to move, they can also prevent that adversary from performing transitions. In turn, this can cause an adversary's military

---

[36] Five basic transitions accompany most operations: 1) transition from movement to attack or defense; 2) transition from attack to defense; 3) transition from defense to attack; 4) transition from an existing form of warfare to a pursuit; and 5) transition from one form of warfare to a retrograde or withdrawal. These should be added to planning priorities, both for an actor's own benefit and for more effectively thwarting an opponent. Reserves are a critical capability for transitions. A reserve's employment is generally tied to one of three options: 1) exploiting tactical or operational success; 2) overcoming an initial failure toward mission accomplishment or attaining an objective; and 3) initiating a pre-identified transition. As with the five basic transitions, adding these three reserve planning considerations to planning priorities will assist a planning team in accounting for reserve employment and its integration with transitions. See the author's "On the Employment of Cavalry," ARMOR 123, no. 1 (Winter 2020), 36-37.

operations to stagnate, make their static formations subject to identification and destruction, and generally increase their cost, pushing the adversary one step closer to culmination and exhaustion.

Principle #5: Information

Information is the final principle of warfare. Information is the data required to make systems operate. In the case of military forces, this system can be referred to as the warfighting system of a state's military force. Without information, a military force can do little more than blindly move about the battlefield and, because of the absence of information from their senior military leaders and policymakers, blunder about doing what they perceive to be in their best intension.

Data can be 'good', or true, relative to the individual or entity reporting the information. Good data generates good information, which is what a networked warfighting system needs to thrive on the battlefield. Thus, generating and maintaining good data is the primary goal of any military force and the state that puts that military force into the field.

Data can be 'bad'. Bad data are facts fraught with holes because the individual or entity reporting the data does not have access to sufficient vantage points to generate a sufficiently accurate picture. Avoiding bad data is paramount for a military force and its state because bad information moving through a warfighting system often leads to suboptimal operations and incomplete battlefield outcomes.

Data can be 'corrupt'. Corrupt data tends to be the result of an adversary's attempt to mislead an adversary by injecting delusive data into a combatant's warfighting system. Like bad data, corrupt data can cause suboptimal operations and incomplete battlefield outcomes, but corrupt data can also mislead a combatant to the point that they conduct incorrect or unneeded military activities.

Data can be 'denied'. Donella Meadowswrites that, "Missing information flows is one of the most common causes of system malfunction."[37] From a self-oriented, defensive position, data denial means that a combatant can prevent the release of data or prevent the observation of their operations such that an adversary cannot depict the observed force's actions, intentions, or capabilities. Further, data denial can also be a threat-focused offensive activity. A combatant can target an adversary's

---

37 Meadows, *Thinking in Systems*, 157.

ability to collect information, whether that is its physical forces, sensors, and networks, to deny data to the adversary.

Data can also be interrupted. Interrupted data differs from denied data in that when interrupted, some data still makes it to its intended receiver, whereas in denied data the data is severed from reaching the receiver. Interrupted data is useful because it can force an adversary into a situation in which they do not possess a sufficient flow of data from which to make predictable decisions, nor address novel situations as they arise.

Lastly, data can be 'temporal', or subject to the impact of time on the data. Within the temporal category, data can move so quickly that it overwhelms the individual or entity attempting to make sense of the data, and thereby, cause the data analysis to be incomplete, and then allow incomplete information to be fed into a combatant's system. Further, within the temporal category, data can move so slow that it does not provide opportune data, and thus, futile information to be fed into their warfighting system.

Ignorance is information's inverse principle of war. If information enables warfighting, then the absence of information – or ignorance – disrupts warfighting. If obtaining, maintaining, and protecting information is vital for an actor, then it must also follow that denying that to one's adversary is of the same critical importance. Thus, the inverse goal of information is to keep an adversary situationally and strategically ignorant, while doing one's best to prevent that from happening to themselves.

Like, the principles of war, the principles and inverse principles of warfare, are not just a disparate collection of words, but rather the reduction of a basic statement for the first order principles militaries must adhere to when engaged in war. The narrative in Box 4.3 is a useful tool.

Further, it is important to take a holistic look at the inverse principles of war too (see Box 4.4). Doing so will provide a better appreciation for what military forces must do and must protect against in war.

Conclusion

To address the challenges of the future of war, Western military thought must expand beyond the confines of engrained institutional thinking. It must periodically question its assumptions and its extant mental models. To keep pace with change, Western military thought must discharge obsolete ideas and concepts, regardless of how uncomfortable doing so might feel. Western military thought must move beyond appeals to authority to

> **Box 4.3: Framework for Thinking about the Principles and Inverse Principles of Warfare**
>
> In war, a combatant must always remain mentally flexible and be prepared on a wide and deep range of education and experience to address situationally unique battlefield situations (pragmaticism). When conducting military operations and activities, a combatant must not fall victim to predictable forms, methods, and timings. Instead, they must do their utmost to remain elusive so that they thereby become harder to identify, target, and destroy (unpredictability). Moreover, the ability to move allows a military force to conduct military operations, reposition forces across the theater of operations, sustain the force – operationally and tactically, and react to changing civilian situations on the battlefield (movement). The inability to move all but obviates a military force's usefulness whatsoever on the battlefield. Transitions are the mechanism by which pragmatic military forces operate unpredictably and react in a self-interested manner to the political-military situations on the ground for operational and / or tactical betterment (transitions). Executed correctly, transitions can bypass the expensive 'start up' costs of a tactical or operational military activity by using a situation's existing conditions to facilitate quickly moving from a successful attack into a deft pursuit, or perhaps from a stalwart defensive operation into a pulverizing counterattack. None of this can happen, however, without information, for information is the lifeblood that animates military operations. Therefore, the pursuit of, obtainment, maintenance, and protection of data and information is, next to movement, the second most important aspect of warfare (information). Operations for information, to maintain information, and to protect information are first-order priorities for all military forces, whether they be state, non-state, or some other form of irregular or non-state actor.

legitimize its guiding ideas. Moreover, Western military thought must not fall victim to emotionally reacting to flashy videos on social media to make claims about fundamental changes in war and warfare. Military thought is far too a serious business to allow emotionally charged reactions drive adaptations in doctrine and force structure.

96 Conflict Realism

> **Box 4.4: A Holistic Look at the Inverse Principles of War**
>
> In war, military forces must refrain from becoming idealistic about any type of warfare, weapon system, or any other thing that dogmatic beliefs could be associated. Idealism clouds a military force's mind to the realities in warfare, which often exceed the bounds of dogmatic beliefs about warfighting. This makes a military force less, not more effective, on the battlefield because they must then wrestle with their gaps in preparedness for the situation at hand (idealism). In a similar vein, an idealistic adversary is preferrable for non-aligned military forces. This is because the idealistic combatant is often predictable. A predictable foe simplifies the problems of incomplete and private information – a predictable foe simply acts one or two ways in any given scenario and is therefore a much more economical problem to solve than an unpredictable adversary (predictability). A military force must therefore do their best to make their adversary predictable, while remaining aware of their own problem of remaining predictable. Movement makes a military force able to operate both pragmatically and unpredictably, whereas the absence of movement capability causes a force to operate lethargically and in an easily identifiable pattern (immobility). Moreover, a military force that is lacking movement capability is more prone to identification, tracking, targeting, and destruction. Considering that exhaustion and the elimination of an enemy combatant's resources is the ways in which wars are won, then making an enemy force immobile is a catalytic event towards battlefield success. Likewise, caution must be rendered towards this concept applied to one's own military force. A force that lacks movement capability has limited utility for military commanders

Box 4.4: A Holistic Look at the Inverse Principles of War, continued

and political leaders. Therefore, when working through force design considerations, force designers must ensure that they do not fail to account for ample movement capability within their military forces. Moreover, a military force should be self-contained and able to move itself. Land forces, as an example, should not be dependent on air or naval forces for movement within, throughout, or across a theater of operations. If transitions are central elements of pragmatic operations built around the fluidity of tactical and operational movement, then protecting the ability to operate in that fashion is a first order principle, yet at the same time, inducing the opposite effect in an adversary is equally important. While the idea of triggering cognitive paralysis is common amongst commenters, that idea overlooks the magnitude of things that must occur for that to happen. However, a more metered approach – preventing transitions – can have a comparable impact, with less cost. Therefore, while a combatant works to protect their ability to conduct transitions, they must actively work to inject suboptimization into their adversary's military operations (paresis). Finally, keeping depriving an adversary of situational and environmental context and denying their ability to communicate forces a combatant into predictable behavior and therefore much easier to identify, target, and destroy (ignorance). At the same time, a force must not allow themselves to become ignorant. In this case, however, preventing ignorance goes beyond the battlefield. To prevent ignorance, a force must embrace diversity and inclusion, or otherwise become idealistic, predictable, and cognitively immobile.

5 On Urban Warfare

As the litany of conflicts in the post-9/11 period illustrate, urban warfare is a critical element of contemporary war. Each potential operating environment presents its participants with unique challenges to overcome, and urban operating environments are no different. Military activities in urban operating environments, or urban warfare, present a unique blend of challenges. Urban environments inhibit swift vehicle movement, and often causes warfare to make a slow, methodical plod, as battles like the 2016-2017 battle of Mosul illustrate. In that battle, the Iraqi security forces would not move against ensconced Islamic State fighters unless they were covered with artillery and air support. The Iraqi's methodical slog through the city created a spidering wave of death and destruction as it moved forward.[1]

Moreover, urban environments provide protection against ground and air-based fires, as well as hide personnel, equipment, and other sinew of warfare. In the Russo-Ukrainian War, both Russia and Ukraine have used civilian infrastructure to conceal troops and military equipment.[2] And in Gaza, Hamas uses hospitals and other civilian infrastructures to protect itself from its adversaries.[3]

Next to the methodical frontal clearance of an urban area, sieges are the most known operation in urban warfare. Further, sieges are an inseparable part of urban warfare. In the post-9/11 period, sieges have occurred in the US-Iraq War, the US-Afghan War, the Syrian Civil War, the Counter-Islamic State wars in Iraq and The Philippines, and the Russo-Ukrainian War. Sieges are not inconsequential, nor the result of bad

---
1 Amos Fox, "The Mosul Study Group and the Lessons of the Battle of Mosul," *Association of the United States Army*, Land Warfare Paper 130 (2020): 5-9.
2 "Russian, Ukrainian Bases Endangering Civilian," *Human Rights Watch*, 21 July 2022, accessed 4 November 2023, available at: https://www.hrw.org/news/2022/07/21/russian-ukrainian-bases-endangering-civilians.
3 John Kirby, *Press Conference on Israel-Gaza War*, 14 November 2023, accessed 15 November 2023, available at: https://youtu.be/nZ8JwsW-nwM?si=lG0Jm2DbbP3VgUAb.

tactics. Sieges are often reflective of a specific situation, coupled with the deterministic impact of urban terrain on military operations. Similar to the situation in Mosul, sieges tend to develop out of one belligerent seeking protection in the urban area, while the other belligerent attempts to manage the situation by encircling their ensconced and bludgeoning and starving them into submission. Russia's siege of Mariupol in 2022 provides a near perfect case example of this point. Russian military forces concentrically defeated Ukrainian defenses outside, then inside the city. As a last-ditch effort to hold the city, Ukrainian forces consolidated in the Azovstal iron and steel plant in late February 2022. Ukraine's 3,000 soldiers were able to fend off Russia's 12,000-man army for 80 days. Nevertheless, Russia's overwhelming strength caused the defenders to surrender on 20 May 2022.[4]

Despite all the great work done on urban warfare to date, there is very little analysis on the relationship between strategy and urban warfare. This chapter examines urban warfare not through the lens of operations or characteristics, but by examining the phenomena through the lens of strategy and conflict realism. Many methods for defining strategy exist, but for the purpose of analysis the *Ends-Ways-Means-Risk* heuristic is a simple, yet helpful tool that helps illustrate urban warfare considerations that might otherwise be overlooked or ignored. Further, in keeping with the tenets of conflict realism, this chapter uses military force as the subject of analysis. Yet, because this chapter examines strategy, military forces are examined in relation to their ability to accomplish their state, or polity's, political-military objectives.

This chapter's analysis suggests that contrary to popular opinion, the defender in urban warfare gains an initial protection advantage, but if its adversary gains its situational bearing, urban operating environments turn into a death trap. Moreover, urban warfare tends to contribute to the proliferation of wars of attrition, regardless of how a state wants to prosecute the battle. Lastly, support for smaller, lighter forces, considering the relevance of urban warfare in nearly every modern conflict, might be incorrect.

This chapter proceeds as follows. First, it engages with 'ends', or a state, military force, or non-state actors' goal within a specific conflict. Lastly, the chapter explores risk's role in strategy and how that consideration shapes

---

4 Michael Schwirtz, "Last Stand at Azovstal: Inside the Siege That Shaped the Ukraine War," *New York Times*, 24 July 2022, accessed 4 November 2023, available at: https://www.nytimes.com/2022/07/24/world/europe/ukraine-war-mariupol-azovstal.html.

how states and non-state actors negotiate their way through urban warfare. The chapter concludes with some thoughts on how strategy can better support policymakers, military professionals, and scholars who must toil on the challenges of urban warfare.

Urban Operations and Strategy

Ends

An 'end' in strategy is a policy or strategic military goal or outcome. Ends evolve as conflict evolves, or so they should. Nevertheless, the ultimate end for any combatant – whether a state or non-state actor – is three-fold.

First, a combatant engages in war to win. Further, winning is not universally defined. As discussed in the preceding chapter, a definition of victory (or winning) is tied to each belligerent and their geopolitical situation, coupled with the status and disposition of their military forces. A belligerent's initial conditions are an important factor, but they do not generate a deterministic pathway that preordains a conflict's outcome.

A correlation of forces and means comparison between Russia and Ukraine in early February 2022 pointed to Russia possessing a significant materiel advantage in almost every measurable category.[5] Angela Dewan characterized the disparity as a David and Goliath situation, in which Ukraine was the smaller and feeble David, and Russia was the imposing Goliath. The disparity in starting, or initial conditions, caused many commenters, policymakers, and military professionals to assume that, if attacked, Kyiv would fall to Russia's military forces within a week, and that most, if not all of Ukraine would revert to the Kremlin's control.[6] However, Ukrainian President Volodymyr Zelenskyy, the Ukrainian armed forces, and the Ukrainian people's stalwart defense against Russia's initial assault in Kyiv, Kharkiv, and other cities demonstrated that initial conditions are not deterministic. Ukraine's collective success in the first desperate days

---

5 Angela Dewan, "Ukraine and Russia's Militaries are David and Goliath. Here's How They Compare," *CNN*, 25 February 2022, accessed 4 November 2022, available at: https://www.cnn.com/2022/02/25/europe/russia-ukraine-military-comparison-intl/index.html.
6 "General Assembly Overwhelmingly Adopts Resolution Demanding Russian Federation Immediately End Illegal Use of Force in Ukraine, Withdraw All Troops," *United Nations*, 2 March 2022, accessed 4 November 2023, available at: https://press.un.org/en/2022/ga12407.doc.htm; "NATO's Response to Russia's Invasion of Ukraine," *NATO*, 6 November 2023, accessed 10 November 2023, available at: https://www.nato.int/cps/en/natohq/topics_192648.htm.

of Russia's invasion, coupled with impressive information operations, and the international community's general displeasure with Russian President Vladimir Putin's aggressive foreign policy toward Ukraine, resulted in a massive wave of financial, materiel, and intelligence support.[7] This dynamic changed the balance between Russia and Ukraine's initial conditions and each belligerent's strategic ends.[8]

Further, a state's unique definition of victory must be viewed as a set of defined conditions, or a win-set. Win-sets are a combatant's codified conditions for victory. Some conditions are transparent and readily available to the international community, whereas other conditions are private information and reserved for a small group of trusted agents. Private information pertinent to a combatants win-set help explain that subject's actions in situations that might not make sense to the outside observer. This also helps explain Putin's seemingly illogical invasion of Ukraine. Putin has private information and win-sets that are not privy to the international community, which makes his actions and that of Russia's appear irrational.

Hamas' incursion into southern Israel on 7 October 2023 is another example of both the ideas of unique definitions of victory and private information. Hamas entered Israel and killed more than 260 people.[9] Hamas' geopolitical and religious interests are well established. Yet, for the curious onlooker, and most of the international community, Hamas' attack seems irrational.[10] Applying the principle of private information one can infer several assumptions to Hamas' known interests. Doing so, thereby allows one to deduce that Hamas possesses external international support, thus increasing their odds for obtaining the conditions outlined within their own unique win-set.

Russia's war in Ukraine is another example of this reality. Russia and Ukraine each have a unique definition of victory. Russia's interest resides

---

7 "Fact Sheet: One Year of Supporting Ukraine," *White House*, 21 February 2023, accessed 4 November 2023, available at: https://www.whitehouse.gov/briefing-room/statements-releases/2023/02/21/fact-sheet-one-year-of-supporting-ukraine/.
8 Gerry Doyle, Anurag Rao, and Vijdan Mohammad Kawoosa, "Shaping the Battlefield: How Weapons from Western Allies are Strengthening Ukraine's Defense," *Reuters*, 10 March 2023, accessed 4 November 2023, available at: https://www.reuters.com/graphics/UKRAINE-CRISIS/ARMS/lgvdkoygnpo/.
9 Patrick Kingsley and Ronen Bergman, "The Secrets Hamas Knew About Israel's Military," *New York Times*, 13 October 2023, accessed 31 October 2023, available at: https://www.nytimes.com/2023/10/13/world/middleeast/hamas-israel-attack-gaza.html.
10 Haroro Ingram and Omar Mohammed, "The Logic of Insanity: Why Groups Like ISIS and Hamas Strategically Court with Self Destruction," *George Washington University Program on Extremism*, 1 November 2023, accessed 5 November 2023, available at: https://extremism.gwu.edu/logic-insanity.

in fracturing Kyiv's sovereignty by taking broad swaths of Ukrainian territory and incorporating that land and its people into the Russian Federation. Kyiv, meanwhile, is interested in retaining its sovereignty over Ukraine's territory and the Ukrainian people. Further, considering the external support provided by the US and other Western states, Ukraine is also interested in reclaiming its lost element Donetsk and Luhansk oblast, as well as Crimea. Private information keeps each belligerent guessing about what their adversary's true goals are within the conflict. This causes the conflict to continue moving forward with no apparent end in sight. This dynamic, in which Russia and Ukraine cycle through seemingly illogical combat operations, is reflective of why wars drift to long wars of attrition.

What's more, modern militaries operate as part of a complex, open and adaptive system. Systems, like any institution, organization, or individual, operate according to the law of 'do no harm'. This is not a new idea. Peter Paret notes that, "For most people the problems posed by war have always been a matter, first of survival, and second of victory."[11] Moreover, J.F.C. Fuller writes that "Self-preservation is the keystone in the arch of war."[12] Alexander Svechin also advocates for the importance of survival in military matters. Svechin states that the first rule of war is to guard oneself against any decisive blows.[13]

Considering survivability's fundamental position within both war and warfare it must be regarded as an unshakable law of war. Strategic actors, whether state of non-state, engaged in war will not intentionally chose to operate in ways or along pathways that lead them to their respective destruction.

Appreciating that war is an adversarial endeavor, in which combatants jockey for dominance, then 'destroy' must be understood as 'survives' inverse equivalent. If Combatant A applies the principle of 'survive' to himself, and he understands that Combatant B is also applying that principle to itself, and to win the former must deny the latter its win-sets, then it must pursue Combatant B's destruction. This idea is not new either. Carl von Clausewitz reminds us that, "So long as I have not overthrown

---

11 Peter Paret, *The Cognitive Challenges of War, Prussia 1806* (Princeton: Princeton University Press, 2009), 129.
12 J.F.C. Fuller, *Generalship, Its Diseases and Their Cure: A Study of the Personal Factor in Command* (Harrisburg, PA: Military Services Publishing Company, 1936), 26.
13 Alexander Svechin, *Strategy* (Minneapolis, MN: East View Information Services, 2004), 248.

my opponent I am bound to fear he may overthrow me. Thus I am not in control: he dictates to me as much as I dictate to him."[14]

Third, war is wasteful. Nevertheless, combatants will not operate in ways or along pathways that accelerate them towards resource exhaustion. Because war is wasteful, and it occurs in a resource-constrained environment, the manipulation of resources – available and potentially available – is a salient pathway to victory or defeat in war and warfare. At the tactical level, depriving a military force of resupply after they have used their on-hand stocks is a proven way to defeat that force. Extrapolating that to the operational and strategic levels is more challenging, but it is still one of the most effective paths to victory.

Ends – analysis
Synthesizing these three concepts – winning, military goals, and survival – the allure of urban operations becomes apparent. Modern military forces often lack the size to operate along vast fronts that contain multiple points of combat. A few exceptions to this rule include the US, Russia, Ukraine, and China. Nonetheless, an aggressor's ability to force their adversary into an urban area centralizes, compresses, and reduces the number of military problems that the aggressor must address. Although the challenges of urban operations are significant, from an operational and strategic level, centralized operations and campaigns simplify the logistics challenges associated with operating multiple battles along a dispersed front. As a result, urban operations remain an attractive option for states and non-state actors seeking to fight and survive on the battlefield.

Further, operating from an urban area increases the military force's chance of survival. Yet, this comes with a caveat. A dataset on post-Cold War sieges – one of urban warfare's defining characteristics today – finds that the defender is successful in only 30 percent of observed cases.[15] On the other hand, the aggressor won 60 percent of the time. The remaining percentage is split amongst ongoing sieges, stalemates, and multiple victors.

Refining the post-Cold War sieges into four categories yields useful findings. In the ten sieges lasting one month or less, the aggressor won 90 percent of the time, while the defender won the remaining ten percent. In the 21 sieges lasting between one and six months in duration, the aggressor

---

14 Carl von Clausewitz, *On War* (Princeton: 1984), 77.
15 *Post-Cold War Siege Dataset_28 October 2023*. Dataset is currently unpublished and retained and maintained by the author. The dataset will be published as part of the author's forthcoming book which is due to be published in 2024.

came out on top 57 percent of the time, while the defender won 33 percent of those sieges. The remaining sieges during that window are split between ongoing sieges and ceasefires.[16]

Further, the data indicates that the aggressor's monopoly on winning sieges begins to change as the duration of the conflict elongates. This trend continues in the 11 sieges lasting between six to 12 months. During that period, the aggressor only won 36 percent of the time, whereas the defender prevailed 55 percent of the time. The remaining nine percent goes to a single stalemate.[17]

A final review of the data finds a so-called sweet spot for the defender. The defender's sweet spot is sieges lasting longer than one month, but less than a year. Although the dataset does not elaborate on causality, the sweet spot presumably emerges because of the synergistic effect of surprise and resources on the operation. Based on the data, a simple theory emerges. Aggressor's win short sieges because they surprise their opponent and bring overwhelming strength to bear simultaneously. The aggressor encircles and defeats the defender before they can regain their bearing. Defeat ostensibly originates from one of two situations. The defender acquiesces because they realize that they are not in position to resist the aggressor or the aggressor might physically destroy the defender's ability to resist before it has the opportunity to establish a suitable urban defense.

Nonetheless, the defender's position stabilizes in sieges occurring between one and 12 months. This results from one of three situations. First, a degree of parity exists between the aggressor and the defender, and by missing the opportunity to defeat the defender, that parity lends victory to the combatant who uses the most prudent and timely operations. Second, the defender activates its bases of power – existing, latent, and potential – to support their operations. Therefore, instead of depleting their personnel, food, and equipment, the defender can parry the aggressor's offensive action. Third, the international community joins the conflict. In sieges lasting longer than a month, for instance, the international community pressures the aggressor to end the siege, or open humanitarian corridors for the besieged population.[18] A besieged actor's best chance of winning

---

16 *Post-Cold War Siege Dataset_28 October 2023.*
17 *Post-Cold War Siege Dataset_28 October 2023.*
18 Bill Hutchinson, "How Humanitarian Corridors Work to Offer Lifeline to Besieged Ukrainians," *ABC News*, 12 April 2022, accessed 4 November 2023, available at: https://abcnews.go.com/International/humanitarian-corridors-work-offer-lifeline-besieged-ukrainians/story?id=84011869.

results from extending a siege to a month, at a minimum, but no longer than a year.[19]

The balance tips back to the aggressor for sieges lasting longer than a year.[20] This is likely because of resources. The basic assumption is an aggressor wins in sieges longer than 12 months because they have the logistical base to outlast the defender. The data also suggests that states are victorious 39 percent of the time when sieges exceed 12 months. Principal-proxy dyads account for a further 17 percent of the wins in this category. Taken collective, states prevail in 56 percent of the sieges that fall within the one-month-to-twelve-month category.[21]

The data indicates that sieges are an inseparable element of urban warfare. Moreover, the analysis shows a trend that contrasts with contemporary assertions regarding urban warfare. Scholars, military professionals, and others suggest that urban environments provide a defender or weaker combatant with a place of refuge from an aggressor or stronger combatant. This might be true, but siege data suggests that unless a defender can elongate the siege into, but not exceed, the one-to-six-month window, their chance of winning a siege is quite low. This finding is specific to a siege, but given the siege's central position in urban warfare, and the absence of any comparable quantifiable information on urban operations, then it is reasonable to assume that urban operations follow a similar pattern. This similarity likely follows both the findings for the relationship between duration and outcomes, as well as the relationship between duration, status (that is, aggressor or defender), and outcomes.

Strategy pertaining to urban operations therefore indicates that they benefit the aggressor most often. As a result, an aggressor should not fear urban warfare. They must either strike with sufficient force and speed to surprise and annihilate the defender before they can respond, or they intentionally extend the duration of the campaign to such a point that the defender can no longer muster the resources required to maintain operations.

To improve one's potential for winning in urban warfare, one must not dither around the margins of force design. Western militaries must increase – not decrease – the size of their forces. This will help Western military forces

---

19 Amos Fox, "Urban Warfare, Sieges, and Israel's Looming Invasion of Gaza," *War on the Rocks*, 27 October 2023, accessed 4 November 2023, available at: https://warontherocks.com/2023/10/urban-warfare-sieges-and-israels-looming-invasion-of-gaza/.
20 Fox, "Urban Warfare, Sieges, and Israel's Looming Invasion of Gaza."
21 *Post-Cold War Siege Dataset_28 October 2023*.

compensate for the destruction and casualties that accompany operating in an urban environment.

Western militaries must also expand their understanding of urban warfare by broadening the concepts and doctrine. To that end, Western militaries must not rely on institutional concept developers and doctrine writers to address the challenges of developing the concepts and doctrine. Extant concept developers and doctrine writers tend to be process-oriented bureaucrats who are not paid to think and generate novel approaches to new problems; but rather, they focus on process and protecting their institutional interests and prerogatives.

The research indicates that Western militaries would be better served identifying the small number of urban warfare experts that exist across friendly states and giving them the lead to develop those concepts and ideas. Doing so will help Western militaries develop innovative DOTMLPF (Doctrine, Organization, Training, Materiel, Leadership and Education, Personnel, Facilities, Policies, and Finance) solutions to enable Western militaries to more effectively and efficiently accomplish their military ends.

Ways

'Ways' refers to the method(s) that a state or non-state actor employs to achieve their ends. Ways are 'means' informed; meaning that if a solution to a problem requires more resources than that belligerent can generate, then that approach or plan is not feasible and is discarded.

Ways are not the product of a state's recalcitrant determination to make one type of warfare fit the situation. Instead, ways reflect a handful of deterministic factors, to include the operating environment, time, chance, and lastly, oneself. George S. Patton makes this point, writing that:

> One does not plan and then try to make circumstances fit those plans. One tries to make plans fit the circumstances. I think the difference between success and failure in high command depends upon the ability, or lack of it, to do just that.[22]

Considering the goal of ways in military strategy and how they must fit the situation, scholars, military professionals, and policymakers must understand a set of key considerations. First, ways must be informed

---

22 George Patton, *War as I Knew It* (New York: Houghton Mifflin, 1995), 92.

by the strategic, operational, and tactical operating environment. Each of these environments bring their own unique set of considerations. The strategic operating environment is a derivative of the governing effect of the international community, International Humanitarian Law, and the strategic adversary. Each of those variables both directly and indirectly shape a combatant's intentions, ends, and ways. Israel's conflict in Gaza helps illustrates this point.

The international community has offered both support and condemnation, depending on the strategic actor.[23] States, international government organizations, and nongovernmental organizations have all urged Israel and the Israeli Defense Forces to adhere to International Humanitarian Law.[24] Israel delayed its invasion of Gaza at the request of the US so that the US could situate air defense systems in the region and to provide more time for accommodations to be made to protect civilians in Gaza.[25]

The operational-level operating environment is the general theater of conflict, or the entire area in which the war is taking place. Considering this variable requires assessing the geographic terrain, to include identifying seaports, airports, and other strategic embarkation and debarkation nodes. This helps a belligerent develop a comprehensive understanding of the theater operating environment's potential impact on military operations. Further, the purpose of this analysis to is generate warfighting approaches that integrate considerations from the strategic operating environment with the constraints and opportunities afford by both a theater's physical terrain and its accessibility by air, sea, and land. Factors such as cyber,

---

[23] "Joint Statement on Israel," *The White House*, 9 October 2023, accessed 4 November 2023, available at: https://www.reuters.com/world/middle-east/israel-bombards-gaza-prepares-invasion-biden-urges-path-two-states-2023-10-25/; Nidal al-Mughrabi and Emily Rose, "Israeli Troops Raid Gaza as Arab Nations Condemn Bombardment," *Reuters*, 26 October 2023, accessed 4 November 2023, available at: https://www.reuters.com/world/middle-east/israel-bombards-gaza-prepares-invasion-biden-urges-path-two-states-2023-10-25.

[24] Andrea Shalal and Kanishka Singh, "Biden, Key Western Leaders Urge Israel to Protect Civilians," *Reuters*, 22 October 2023, accessed 4 November 2023, available at: https://www.reuters.com/world/biden-holds-call-with-key-western-allies-pope-discuss-israel-hamas-war-2023-10-22; "General Assembly Adopts Resolution Calling for Immediate, Sustained Humanitarian Truce Leading to Cessation of Hostilities Between Israel, Hamas," *United Nations Tenth Emergency Special Session*, 40th and 41st Meetings, 27 October 2023, accessed 4 November 2023, available at: https://press.un.org/en/2023/ga12548.doc.htm.

[25] Dion Nissenbaum, Gordon Lubold, Don Lieber, and Omar Abdel-Baqui, "Israel Agrees to US Request to Delay Invasion of Gaza," *Wall Street Journal*, 25 October 2023, accessed 4 November 2023, available at: https://www.wsj.com/world/middle-east/israel-battles-on-multiple-fronts-as-conflict-risks-spreading-a5e537ec.

information, and other lesser, but important, variables are also considered here. Moreover, further operational level considerations include examining a theater for the relationship between road networks to (and from) urban areas, large bridging, the density of urban areas, and the ratio of urban areas to less restrictive terrain within a specific theater's context. Patton is helpful here too. He writes, "Surely the greatest study of war is the road net."[26]

Continuing along this line of causality, Patton asserts that high-level military commanders are not concerned with how to defeat an adversary. That concern, according to Patton, is the prerogative of tactical commanders. Operational and strategic military commanders are concerned with where to defeat their adversary. Patton continues, stating that, "The where is learned from a careful study of road, rail, and river maps."[27]

Considerations at this level also require assessing the theater's geography in relation to potential types of military activities to determine time, resource, and consumption rate factors needed to support operations. Theater level operating environment considerations allow policymakers and military leaders to craft military activities that align military forces with theater-specific considerations pursuant to political and strategic military objectives. The US's invasion of Iraq helps make this point tangible. The US invaded Iraq on the flawed belief that once Saddam Hussein's regime fell that the Iraqi people would welcome their forces as a liberating force for good.[28] Poor policy decisions involving Paul Bremer and the Coalition Provisional Authority's decision to de-Ba'athify Iraq and to disband the Iraqi Army, coupled with insufficient troops on the ground (the product of an egregious misread of the strategic and theater operating environment), institutional proclivity (ahead of military necessity) and an inefficacious military doctrine worked in unison to derail the US's policy and strategic military objectives.[29] The derailment caused the US to develop a subsequent policy and military strategy (i.e., counterinsurgency, or COIN) to address the problems generated by its initial strategy.[30] In turn, the US's initial bad

---

26 Patton, *War as I Knew It*, 92.
27 Patton, *War as I Knew It*, 354.
28 Michael Gordon and Bernard Trainor, *Cobra II: The Inside Story of the Invasion and Occupation of Iraq* (New York: Pantheon Books, 2008), 15.
29 US Secretary of Defense Donald Rumsfeld's advocacy of 'transformation' and making US forces small, faster, and sleeker ran in direct contrast to the military necessities for fighting a war in a country dominated by large cities in which most of state's population resided. See Paul Light, "Rumsfeld's Revolution at Defense," *Brookings Institute*, Policy Brief #142 (2005): 1-8.
30 Joel Rayburn and Frank Sobchak, *The US Army in the Iraq War – Volume 1: Invasion – Insurgency – Civil War, 2003-2006* (Carlisle Barracks, PA: US Army War College Press, 2019), 281-303.

strategy – in which its ways, means, and risk evaluations proved incorrect from the start – resulted in Washington's inability to substantially attain its policy goals. In a somewhat fitting end to the war in Iraq, the US Army's own official history, published after several delays, asserts that Iran was the real winner of the Iraq War.[31]

The tactical operating environment is less important when developing strategy because if a decent operational level operating environment assessment is conducted, most of the significant tactical level considerations will have been identified. Nonetheless, tactical considerations help add necessary detail to how a theater's terrain impacts potential operations.

Second, time is a vital element of ways. Speaking of the importance of time in war and warfare, Napoleon Bonaparte stated "I may lose ground, but I shall never lose a minute. Ground, we may recover; time, never."[32] Fuller writes that, "Time is an all-embracing condition…it must be reckoned in minutes, and not only from a military point of view, but from an economic one as well."[33] Olivier Schmitt also notes, in war, "The perception of 'time' changes according to the social-political context." Andrew Carr makes a similar point, stating "Time is an ordering process," and that, "Political and military leaders thus must recognize that their assumptions about the order of events shapes how they interpret the available information."[34]

Time's deterministic impact on every aspect of war and warfare is what makes it germane to the strategy of military activities. If time is short, meaning that the strategic, domestic, or military situation requires a quick military operation, then that limits what a state can do. For instance, if a state's interests are one of unprovoked territorial acquisition, like Russia's 2014 desire to annex Ukraine's Crimean and Donbas territories, the state must move to avoid the condemnation of the international community and the rebuke of a disinterested domestic audience. As Dan Altman notes, this type of situation is often where *fait accomplis* emerge.[35] Unlimited time resides on the other end of the spectrum. The US's unbounded timeframe

---

31 Jeane Godfroy, James Powell, Matthew Morton, and Matthew Zais, *US Army in the Iraq War – Volume 2: Surge and Withdrawal, 2007-2011* (Carlisle Barracks, PA: US Army War College Press, 2019), 569-586.
32 David Chandler, *Napoleon* (London: Pen and Sword Publishers, 2000), 168.
33 J.F.C. Fuller, *The Foundations of the Science of War* (London: Hutchinson and Company, 1926), 180.
34 Andrew Carr, "It's About Time: Strategy and Temporal Phenomena," *Journal of Strategic Studies* Vol. 44, no. 3 (2021): 311. DOI: 10.1080/01402390.2018.1529569
35 Dan Altman, "By Fait Accompli, Not Coercion: How States Wrest Territory from Their Adversaries," *International Studies Quarterly*, Vol. 61 (2017): 882-884.

for its punitive campaign in Afghanistan illustrates how strategy evolves to its time considerations.

The same holds time dynamic true for urban operations. Some urban operations, like the US's siege of Fallujah in the winter of 2004, were influenced by an international community that was sensitive to the plight of the citizens of Fallujah, and therefore applied pressure on US policymakers and strategists to make the operation quick and careful to adhere to International Humanitarian Law.[36]

At the operational and tactical levels, time is often dictated, not directed. By that it must be understood that geography, adversary's military activities, and that adversary's ability to activate bases of power influence a conflict's duration. For example, if the operating environment is dominated by terrain that inhibits movement, then the time needed to accomplish military objectives increases. If road networks and water crossings are limited, the same holds true. As Russia advanced towards Kyiv in February 2022, the Ukrainian military blew the Kakhovka Dam, flooding the main arteries leading from southern Belarus to Kyiv.[37] Doing so caused the Russian advance toward Kyiv to stall.[38] In fact, eliminating access to these roads slowed Russia's operations to the point that Vladimir Putin called off the advance of Kyiv, and ordered a redirect of Russian military forces towards the Donbas sector.[39]

Chance is the third factor that impacts ways. Chance is the impact of randomness and suboptimization on military operations. Randomness is the effect of any unforeseen or unplanned event on a specific situation. Suboptimization is the impact of inherent inefficiencies on military operations. Moreover, suboptimization is the gradual and perpetual towards disorder. Militaries constantly work to combat suboptimization. Training is one way in which policymakers and military leaders attempt to overcome chance and suboptimization. Training provides military forces and the individual soldiers that comprise any military the opportunity to

---

36 Rayburn and Sobchak, *The US Army in the Iraq War – Volume 1*, 344-351.
37 "Ukraine Blew Up a dam to Stop the Russian Advance on Kyiv, Some Homes Remain Flooded," *Reuters*, 28 May 2022, accessed 4 November 2023, available at: https://www.reuters.com/world/europe/months-after-dam-destroyed-stop-russian-advance-parts-village-still-flooded-2022-05-29/.
38 Dmytro Dzhulay, "Revealed: How Ukraine Blew Up a Dam to Save Kyiv," *RadioFreeEurope/RadioLiberty*, 26 February 2023, accessed 4 November 2023, available at: https://www.rferl.org/a/ukraine-russia-moshchun-irpin-kyiv-war-battle/32286263.html
39 Robert Burns, "Russia's Failure to Take Down Kyiv was a Defeat for the Ages," *Associated Press*, 7 April 2022, accessed 4 November 2023, available at: https://apnews.com/article/russia-ukraine-war-battle-for-kyiv-dc559574ce9f6683668fa221af2d5340.

experience a military event or activity in a safe, learning environment before having to deal with that situation in an unsafe, unforgiving environment in which every bullet and artillery round is real.

Institutional or personal preference is the last consideration. This is because the other elements provide situational variance regarding their respective deterministic impact on operations. For example, a group of policymakers might want to hasten a conflict's end. Yet, the problem that they are attempting to address through the use of force is the eradication of a hostile military force in a theater of conflict dominated by urban areas and thick forests. The impact of the physical terrain, coupled with the military objective's own proclivity to avoid existential crisis, means that the conflict will take much longer than what the policymakers would like.

Looking at this argument from another level of examination, what if, in the short scenario provided above, the policymakers still direct a 'short' conflict and that they want to limit civilian casualties and collateral damage during the process. That actor's military commanders then state that they want to conduct a lightning quick campaign of maneuver that seeks to avoid urban warfare and defeat the adversary without having much impact on the civilian population. The physical environment and the threat's interest in survival, however, will cause the conflict to be flipped on its head. An adversary would likely benefit from seeking refuge in the urban area, while using the thick wooded area outside its cities to slow its adversary and cause it to move along predictable routes and force it to fight its way into the urban area.

At the tactical level, given the same scenario, what if a small tactical unit finds itself on the cusp of attacking into urban? This situation was one that neither the policymakers nor the senior military commanders wanted, but the adversary's actions, coupled with the potential benefit that dense urban terrain provides for a defending force, resulted in this outcome. The tactical military commander – steeped in the language and mental models of their respective military institution – also states that they want to conduct a battle of maneuver (in the urban area) and that they want to limit the impact of their military activities on the immediate population and infrastructure. Yet, once hostilities commence, the adversary contracts into the city's recesses and operates amongst the people in questionable ways, to include using them as human shields and relocating local civilians against their will to the areas in which combat is occurring.

The scenario laid out above is not too far from reality. The scenario mirrors what occurred during the US-Iraqi Mosul campaign to eliminate the

Islamic State in 2016-2017. Iraq's open deserts did not provide the Islamic State any military advantage, so they operated within urban areas. Despite the US-Iraqi dyad's best efforts, they were unable to separate the Islamic State from the civilian population sufficient to limit civilian casualties and collateral damage. As a result, cities like Ramadi and Mosul were razed during the process of annihilating the Islamic State. Tracing the footsteps back to the policymakers in this example finds that a leader's preferential method of warfare had almost no impact on how their forces were forced to fight. Therefore, it is important for policymakers, scholars, and military professionals to understand that dynamic. As a result, policymakers and military professionals must integrate non-preferential forms of operating and non-preferential operating environments into how they physically and mentally prepare their forces for war.

Ways – analysis

The operating environment, time, chance, and an opponent's interest in self-preservation and survival all work together to undercut the best laid plans. What does this mean as we examine strategy and urban warfare? It means that all things being equal, a weaker military force will not intentionally meet a strong military force in terrain that provides advantage to the stronger force. In fact, the opposite rings true. If provided time and the means to do so, a weaker combatant will move from open terrain and into urban areas because of the benefit provided by urban infrastructure. Given the fact that military forces are in a period of shrinking size, and armed reconnaissance-strike drones and long-range fires are now offsetting the gaps created by the use of larger land forces, it then makes sense that once a stronger force-weaker force dyad appears, that weaker forces will operate from a city to try and achieve some degree of parity and increase its odds of survival and military success. Consequently, urban warfare will likely play an important role in war because it provides a combatant with game-changing potential.

Means

'Means' are perhaps the simplest element of strategy to examine within the context of urban warfare. Means are the resources required to animate any military activity. Most often, means are considered as the military forces, to include their personnel and equipment, required to engage in war. At the policy and strategic military levels, means include many broader topics such as economic considerations, strategic lift to move forces across the globe, or within (and between) military theaters.

Protection of means, like survival and self-preservation, is a first order principle for any belligerent engaged in war. Urban areas, more so than any other type of operating environment, provide a military force with the best protection for their means. Military forces that possess sophisticated reconnaissance, strike, long-range fires, and professional land forces want to engage adversaries in terrain that favors those advantages. Conversely, it is suicidal for a military force to engage with a military of equal or greater strength without attempting to do so from an area that protects its means and offset the belligerent's advantages.

Means also carries a telescopic aspect with it. By that, means represent different things at different levels of analysis. At the tactical level, means are thought of as accessible warfighting resources. At the operational level, however, means are one of the primary factors upon which exhaustion balances. Exhaustion is the state in which a belligerent can no longer sustain their rate of resource consumption in relation to the tempo or physical extent of military operations. Nazi Germany's operations in Stalingrad are instructive here.

By December 1942, Soviet military forces had surrounded General Friedrich Paulus' 6th Army (some 270,000 soldiers) in the city of Stalingrad.[40] In doing so, Soviet forces cut Paulus and the 6th Army's ability to attend to their means by the replenishment of supplies.[41] Antony Beevor notes that by January 1943, Nazi Germany's political leadership came to the realization that 6th Army was exhausted and that they were able to do little more than feeble localized defensive operations.[42] The once powerful German war machine, short on military and economic means, was unable to provide Paulus with the resources he needed to keep his quarter-million man army afloat.[43] Paulus and his exhausted army fought on through early February 1943. When culmination set in, Paulus surrendered himself and the 6th Army.[44]

---

40 David Glantz and Jonathon House, *To the Gates of Stalingrad: Soviet-German Combat Operations, April-August 1942* (Lawrence, KS: University Press of Kansas, 2009), 486.
41 Antony Beevor, *Stalingrad, The Fateful Siege: 1942-1943* (New York: Penguin Books, 1998), 308-311.
42 Beevor, Stalingrad, 308-311.
43 Beevor, Stalingrad, 308-311.
44 Robert Citino, *The Death of the Wehrmacht: The German Campaigns of 1942* (Lawrence, KS: University Press of Kansas, 2022), 157.

Means – analysis

Considering the preceding ideas in the context of urban warfare is important because it provides a handful of important findings. First, because means are a combatant's lifeblood, protection of one's means is a first order principle of warfare, regardless of the operating environment's physical characterization. Moreover, weighing scholar Anthony King's research on the relationship between the diminishing size of state military forces and the reciprocal increase in urban warfare, finds that small land forces operating in open, unobstructed terrain are more prone to identification and destruction. Considering the requirement to protect means, and the inadequacies of modern forces structures in unobstructed terrain, it is simple to deduce that urban warfare will remain a significant consideration for Western military forces. It is not a stretch to actually see the phenomenon accelerate as states look to further reduce the size of their forces and replace that loss of physical mass with long range precision strike.

Long range precision strike, however, requires excellent reconnaissance to work. Left unchecked, the proliferation of sensors – an evolved form of reconnaissance – will make battlefields prone to precision strike and long-range precision fires. Moreover, the relationship between transparency and strike will require active measures to protect tactical and operational means. Further, the protection of tactical and operational means helps prevent a combatant from expending more resources than its strategic base can support. As a result, the assumption must be made that urban environments will become even more important than they are today because of their ability to both hide and protect people, equipment, and other means of warfare.

This situation should cause the onlooker to take pause and question some key ideas in military discourse today. First, is the evolution to smaller land forces (and militaries, in general) a deterministic pathway that cannot be escaped? Second, are small, lighter land forces the answer to the problems of contemporary, and increasingly, future war?

The data is clear on these questions – small land forces are easier to annihilate, easier to identify through their pattern of sustainment support, and due to their small size, they are far less capable of addressing the challenges of land warfare, and urban warfare in particular. It is therefore reasonable to suggest that small, lighter land forces might provide some advantages in deployability and the ability to hide within urban operating environments, but neither of those advantages seem to outweigh the advantages of larger, heavier land forces.

Further, small, lighter land forces have yet to prove themselves in large-scale combat operations. The US military's initial 'light footprint' approach in Afghanistan and Iraq provided some small victories early in each of those conflicts. These small victories were presumably due more to the resultant of surprise and the localized overmatch the US was able to achieve, and less on the effectiveness of 'light footprint' strategy. In due time, the 'light footprint' strategy caused more problems than victories delivered.[45] Moreover, those problems proved much more strategically disruptive and long-lasting than did the impact of the early victories.[46] Once military leaders acknowledged the 'light footprint's' failure, they pleaded with policymakers to loosen the restraints on troop deployments in both conflicts. Both Afghanistan and Iraq required troop surges to account for the 'light footprints' failures.[47] The failures of strategy associated with the light foot print has caused many other commenters to suggest that the approach is a paradoxical fool's errand.[48] Though these conflicts are not a direct facsimile for conflicts in which large-scale combat operations between industrialized states occur, it is reasonable to assume that 'light footprint' strategies will be disastrous in the future, and should therefore be avoided, despite the financial cost.

Risk

'Risk' is a diverse idea. Regarding strategy, and specifically to strategy pertaining to urban warfare, risk should be understood as the trade-off between what is possible with what it costs to generate that outcome. Costs are more than financial considerations. Costs also include things such as civilian casualties, collateral damage, the impact on personnel (for example, will losses be worth the cost of expenditure), the impact on weapon systems and munitions stockpiles, and a host of other considerations. Thomas Schelling provides a useful way to think about risk. Schelling writes, "The questions that do arise involve degrees of risk – what risk is worth taking,

---

45 Bad Stapleton, "The Problem with the Light Footprint: Shifting Tactics in Lieu of Strategy," *CATO Institute*, 7 June 2016, accessed 4 November 2023, available at: https://www.cato.org/policy-analysis/problem-light-footprint-shifting-tactics-lieu-strategy.
46 Stapleton, "The Problem with the Light Footprint."
47 George Bush, "The Surge of Troops in Iraq," *PBS*, 4 May 2020, accessed 4 November 2023, available at: https://www.pbs.org/video/george-w-bush-troops-surge-iraq/; Barack Obama, "The Way Forward in Afghanistan," *Obama White House Archives*, accessed 4 November 2023, available at: https://obamawhitehouse.archives.gov/issues/defense/afghanistan.
48 Stapleton, "The Problem with the Light Footprint."

and how to evaluate risk involved in a course of action…it adds an entire dimension to military relations: the manipulation of risk."[49] Olivier Schmitt builds upon Schelling's idea, as well as many other scholars, by writing that in the post-Cold War era war is the art of "risk management."[50]

The urban operating environment is a slippery slope. A state or non-state actor choosing to operate in an urban area theoretically decreases the risk associated with using their own forces. In an applied situation, however, the decreased cost to one's own troops also comes at the cost of the inherent civilian casualties and collateral damage that accompanies urban warfare. Moreover, the use of precision guided munitions is often cited as a way in which Western militaries limit civilian casualties, collateral damage, and adhere to International Humanitarian Law. Unfortunately, historical analysis of post-Cold War precision strike-laden warfare tells another story.[51] Warfighting strategies underpinned by precision strike, contrary to the laudatory narratives, prove just as destructive, if not more destructive, than conflicts without precision strike.[52]

Proxy force employ is another byproduct of risk management in war as states look to offset and ameliorate the costs associated with war that have increasingly sought to use proxies to bring that vision to reality. Proxy wars – in which multiple proxy strategies are being utilized by states to support their self-interested policy objectives – are on the rise since the end of the Cold War.[53]

Moreover, the impact of proxy utilization on an operating environment is often overlooked in proxy war and urban warfare scholarship. Proxies are often third parties who are not concerned with International Humanitarian Law or the negative consequences of their actions on their principal's policy objectives or strategic military objectives. A state's use of proxies often exacerbates collateral damage and the loss of civilian life in urban operating environments. Moreover, because proxies are not a state's own military force, states haphazardly throw their proxies into situations in which annihilation is omnipresent. Jack Watling and Nick Reynolds highlight this point by articulating Russia's use of Wagner Group proxy

---

49 Thomas Schelling, *Arms and Influence* (New Haven, CT: Yale University Press, 1966), 94.
50 Olivier Schmitt, "Wartime Paradigms and the Future of Western Military Powers," *International Affairs*, Vol. 96, no. 2 (2020): 5.
51 Amos Fox, "Precision Gone Wild," *RUSI Journal* (Forthcoming).
52 Amos Fox, "Precision Gone Wild."
53 Andrew Mumford, "Proxy Warfare and the Future of Conflict," *RUSI Journal* Vol. 158, no. 2 (2013): 42-43.

forces in Bakhmut as 'meatgrinder' tactics.[54] The same pattern is present in Russia's employment of Wagner during the battle for Mariupol.[55]

This pattern was also seen during the US's use of Iraqi Security Forces as proxy land forces during Operation Inherent Resolve. The term 'meatgrinder' might well do justice in characterizing how the US encouraged the Iraqi Security Forces to operate during the battle of Mosul. Watling and Reynolds are again helpful here. They point to the fact that the Iraqi's Counter Terrorism Service, or CTS, was often the spearhead of operations in Mosul, which resulted in a casualty rate greater than 100 percent for battalion commanders during the siege of Mosul.[56]

Although challenging to prove, another hypothesis regarding proxy force employment is that the strategy extends, or elongates, a conflict. Similar to a state's willingness to accept 'meatgrinder' tactics when proxy forces are their predominant land force, states appear less interested in accelerating a conflict to a decisive political outcome in conflicts dominated by proxy strategies. The same logic carries over from the 'meatgrinder' scenario. States' domestic and strategic risk tolerance is greater when proxy forces are used because the negative consequences that often accompany a long war – to include a deluge of that state's uniformed troops coming home in body bags – goes unnoticed by that state's domestic audience. Likewise, strategic concerns are often brushed aside because the operations or outcomes in question are 'someone else' or 'unruly partners', and not the actions of the state itself. Conflict elongation is the nature drift that each of these considerations embodies.

What about the technology's potential impact related to risk and urban warfare? Many commenters suggest that robots, autonomous systems, artificial intelligence, and machine learning will allow states to overcome the challenges of urban warfare. Precision guided munitions, both armed and unarmed reconnaissance drones, and sophisticated sensors have not yet been proven to be able to sidestep the inherent advantages associated with operating in urban environments. Technology might be able to circumvent the urban operating environment's ability to hide combatants amongst

---
54 Jack Watling and Nick Reynolds, *Meatgrinder: Russian Tactics in the Second Year of Its Invasion of Ukraine* (London: RUSI, 2023), 3-8.
55 Mike Eckel, "The Bakhmut Meat Grinder: Russian Troops Are Pummeling This Donbas City. It's Unclear Why," *RadioFreeEurope/RadioLiberty*, 13 December 2022, accessed 4 November 2023, available at: https://www.rferl.org/a/ukraine-bakhmut-russia-assault-invasion-analysis/32174980.html.
56 Jack Watling and Nick Reynolds, *War by Others' Means: Developing Effective Partner Force Capacity Building* (London: RUSI, 2021), 23-26.

and within infrastructure and civilian populations. However, the cost of developing that technology might surpass the threshold of acceptability for even the most financially robust states. The end state nonetheless remains to be seen.

To summarize this section, risk is, and will remain, a significant strategic consideration for policymakers and military professionals forced to address the reality of urban warfare. Neither proxy strategies, nor technological innovation have made significant progress in decreasing the general risk associated with urban warfare. As policymakers and military professionals look to address urban warfare in contemporary and future war, risk should be the first among peers when developing strategy.

Principles of Urban Warfare

As discussed earlier in this book, the principles of war are a worldly idea; they are common property and not the proprietary or intellectual property of a specific organization, theorist, or institution. Further, principles of war – or warfare – should be at the heart of a military force's method of operation. If principles are not embedded in operations, they are then not principles. Moreover, if a method of military operations does not adequately account for their principles of war – or warfare – then those military operations are fundamentally flawed.

Within the broad span of principles of war and principles of warfare, some operating environments have such a deterministic effect on operations that the environment possesses its own set of principles. Urban areas are one such operating environment that requires its own set of principles of warfare. The principles of urban operations follow:

- **Principle 1: Survive.** Survival should be a first order principle any way you slice both the theory and practice of war and warfare. The ideas' absence presumably reflects an underdeveloped appreciation for strategic ideas pertinent to a state's ability to wage war at the cost of more tactically minded principles.
- **Principle 2: Urban operations are inseparable from attrition.** Urban terrain undercuts the advantage of standoff warfare and the ability to maximize distance between two combatants. Forces required to fight in urban environments must therefore be built for the rigors and destruction of urban environments. Further, Western militaries must begin the uncomfortable move of distancing themselves from

fanciful maneuver-centric doctrine and accept destruction-based, and positional, warfighting norms within their respective lexicon. Western nations must also examine their industrial base, and its linkage with the warfighting requirements of forces in attritional wars and modify the network to not quickly bankrupt war stocks once fighting commences.

- **Principle 3: Polarity.** Borrowing again from Clausewitz, it is important to that, "In a battle each side aims at victory; that is a case of true polarity, since the victory of one side excludes the victory of the other."[57] As a result of polarity, Western military thinking must accept that advantages in urban battle are fleeting, because adversaries will find ways to offset strength and equipment asymmetry, just as Ukraine has done by weaponizing social media. Ukraine's use of social media, from the outset of the invasion, is a case study in how information operations can assist in garnering support, and growing one's base of power, in the face of odds that upon initial look are insurmountable.

- **Principle 4: Ground Lines of Communication (Still) Matter.** As Ukraine's urban battles attest, ground lines of communication still matter. Ukraine's southern seaboard, for example, is an amalgamation of cities linked by a dense network of highways and railways. For an army like that of Russia, which relies on bulk supply distribution via rail, and micro-distribution to forward fighting locations with trucks, the rail and highway systems, and their major points of convergence, are critically important. As a result, when forecasting potential points of combat, Western militaries must assess the combatant's logistics distribution system. If a combatant uses a bulk distribution system based on land movement, expect combat in and around areas in which railways and highways converge and diverge. If a combatant uses palletized distribution, expect combat to focus heavily on airports. In both instances, modern industrial warfare is dependent on the logistics network, and logistics hubs almost always reside in urban areas.

- **Principle 5: Proximity and Density.** The degree of civilian casualties and collateral damage corresponds to proximity, and density, of civilians and civilian infrastructure on a battlefield. Therefore,

---

57 Clausewitz, *On War*, 83.

if engagements, battles, or campaigns are being fought in urban areas, then the civilian casualties, displaced persons, and collateral damage will be higher than if fought elsewhere. This principle applies regardless of the tactics or munitions (for example, ballistic or precision) used on the battlefield.

- **Principle 6: Precision Paradox.** In urban warfare, the precision paradox ensures that despite the use of precision munitions, battles will result in wide-scale destruction of civilian infrastructure and the loss of life.[58] Additionally, in urban warfare, which is dominated by a never-ending number of hiding places for an absconding combatant, the use of munitions will remain high.[59] As a result, Western militaries must discontinue the practice of thinking and speaking of precision munitions as a panacea, and as a tool that makes war less horrendous. Instead, they must account for the suboptimal effects that inevitably accompany the use of precision weaponry.

- **Principle 7: Sieges.** The siege of Mariupol, and specifically, the Azovstal industrial plant, highlight the recurrent character of sieges in modern urban warfare. In Ukraine's ongoing war with Russia, which bore witness to significant sieges at Ilovaisk, Luhansk Airport, Donetsk Airport, and Debal'tseve in 2014-2015.[60] Due to urban environment's security from observation and fire strikes, as well as their connection to needed logistics infrastructure, such as railways, highways, and airports, urban sieges will remain a fixture in industrial, wars of attrition.[61] In light of this fact, Western militaries must develop doctrine, organization, and training solutions to account for the siege's absence in their collective thinking.

- **Principle 8: Mobility.** Remaining mobile in urban warfare is paramount for both survival and offensive action. Mobility allows a combatant to remain elusive and hard to kill.[62] Simultaneously, mobility allows an actor to maintain the ability to move, strike,

---

58 Amos Fox, "The Precision Paradox and the Myths of Precision Strike in Modern Armed Conflict, *RUSI Journal* (2024): 4-6.
59 Joshua Andersen, "The Paradox of Precision and the Weapons Review Regime," *Philosophical Journal of Conflict and Violence* Vol. 1 (2017): 17.
60 Amos Fox, "On Sieges," *RUSI Journal* Vol. 166, no. 2 (2021): 2-8. DOI: 10.1080/03071847.2021.1924077.
61 Fox, "On Sieges," 10-11.
62 J.F.C. Fuller, The Reformation of War (London: Hutchinson and Company, 1923), 4.

and defend on the move, which forces the other combatant to have to account for more variables than it would against a static, or fixed, enemy. To be sure, Fuller notes that "Liberty of movement is the basis of liberty of action, which is a compound formed out of superiority in the elements of war."[63] Considering polarity, offensive and defensive action in urban areas must therefore seek to eliminate an adversary's capacity for mobility Denying an enemy movement in an urban area makes it exponentially easier to fix the combatant in place, encircle the combatant, and concentrically eliminate the combatant from the battlefield.

- **Principle 9: Base(s) of Power.** To succeed in urban battles, especially in siege or encirclement situations, a combatant must possess a base, or multiple bases, of power. Accounting for polarity in warfare, strong initial conditions only get a combatant so far. Combatants must activate latent bases of power, and mobilize external bases of power, to maintain pace urban battles of attrition and position. As Ukraine demonstrates, the ability to mobilize external bases of power – such as garnering support from the US, the EU, and NATO members, for instance – has allowed it to overcome correlation of forces and means comparisons that indicated that without support, Ukraine's forces and weapons stockpiles were not long for the world.

- **Principle 10: Center of Gravity (COGs) do not exist.** As the loss of Russian General Officers and the destruction of field armies demonstrates, modern wars against actors with strategic depth in resources, viable ground lines of communication illustrate, and robust force structure, COGs are an antiquated relic of Napoleon interpretations of war and warfare. COGs reflect a mechanistic character of warfare in which heads of state led their armies on the field of battle, thereby creating a tight coupling between the policymaker and the tactical command of forces.[64] Further, COGs made sense when armies, industrial bases, and sustainment networks were less productive, less connected, and not able to

---

63 Fuller, *The Reformation of War*, 4.
64 Amos Fox and Thomas Kopsch, "Moving Beyond Mechanical Metaphors: Debunking the Applicability of Centers of Gravity in 21st Century Warfare, *Strategy Bridge*, 2 June 2017, accessed 5 November 2022, https://thestrategybridge.org/the-bridge/2017/6/2/moving-beyond-mechanical-metaphors-debunking-the-applicability-of-centers-of-gravity-in-21st-century-warfare.

rapidly communicate between one another. In this mechanistic model, eliminating an enemy army from the field of battle had a direct psychological impact on the head of state. Today's militaries, however, are not prone to shock and destruction like the armies of yore. Today's militaries consist of redundant communication and sustainment networks that generally account for losses in ways that negate the continued utility of COG thinking. Therefore, Western militaries must discontinue the practice of placing a premium on COG analysis, and instead focus that attention on systems and network-centric thinking.

To tie these principles into a cogent and unified idea, the following problem statement is provided to allow policymakers, military professionals, and scholars a tool to use when thinking about how to approach or address urban operations.

Problem Statement: *Military operations often occur in urban areas because combatants seek refuge beneath the protective structure of urban terrain (survive). Further, military forces often elect to fight from positions of relative strength in urban areas because the urban terrain offsets an adversary's inherent strengths and advantages, and potentially enhances the adversary's weaknesses (polarity). Urban environments allow combatants to use or abuse bases of power. If a combatant can maintain their bases of power, they capitalize on the benefits of operating from an urban area; however, if their bases of power are eliminated, then they are fighting on borrowed time (bases of power). Supply lines of all types, but specifically ground supply lines are critical for a military force operating from an urban area (ground lines of communication). They are important because they keep the combatants bases of power properly tuned, and thus allow the combatant to continue fighting. Therefore, it is imperative to protect ground supply lines if one is operating from an urban area, but it is equally important to severe those lines if one is operating from outside the urban area and focused on the enemy inside the city. Because of the challenge-response cycle that occurs between combatants in an urban environment, those areas affected by the fighting often feel high levels of death, destruction, and disease (proximity and attrition). Precision strike strategies only go so far in urban areas – a strike might be accurate, but it can be equally ineffective, by not eliminating its intended target, or counterproductive by accidently killing innocent bystanders (precision paradox). Regardless of how precise a strike might be, the explosion of ordinance in tightly packed spaces, such as a city, creates unintended consequences that must be appreciated ahead of time. Further, because of the protective character of urban operations, coupled with the fact that military forces are open, dynamic systems that seek to survive, centers of gravity are of little concern (centers of gravity). Nonetheless, battles of position and encirclement often accompany operations in urban areas (sieges). The goal of these battles is to encircle an adversary within the city, cut their access to resources, and slowly cause systemic collapse and ultimately systemic exhaustion across the breadth and depth of the military force trapped in the urban area. Denying an adversary's ability to move within, to and from, and outside an urban area is critical for the aggressor because doing so prevents the besieged combatant's ability to refresh and maintain their warfighting capacity (mobility).*

Conclusion

Urban warfare presents a unique challenge for policymakers, military professionals, and scholars. Using the *Ends-Ways-Means-Risk* heuristic to examine those challenges is a useful thought exercise because it dissects the problem from a combined standpoint, utilizing theoretical ideas and balancing those with the realities that urban operating environments present. The ways of urban warfare, today, and for the foreseeable future, remain limited. There are only so many ways in which to address a military problem within an urban environment. Moreover, resources both bind a combatant's potential ways, yet afford the belligerent a range of military options, provided the assets to unlock those options are on one's resource menu, or readily available through external supporters.

Ends, like ways, are bound by a state's means. If a state and its military force does not possess the resources required to generate a preferential outcome in urban operating environments, then that outcome is not feasible, and must therefore be discarded. Moreover, one's ends must also account for a belligerent force's military operations. If an adversary is operating in such a way that makes one's policy or strategic military ends out of reach, then that too must be accounted for during the strategy process at both policy and military levels.

Risk, however, is the most paradoxical strategic element. The methods to decrease the risk to one's own forces and to unlock the ends of a state's policy often create other significant challenges. Those challenges tend to include creating high casualties amongst involved military forces, high rates of civilian casualties and collateral damage, and conflict elongation. Conflict elongation, by virtue of expanding the duration of a conflict, also increases military and civilian casualties, and collateral damage. Taking a step back from the individual outcomes in urban warfare – increased military and civilian casualties, collateral damage, and a conflict's longer duration – finds that war within urban areas tend toward being defined as wars of attrition.

As a result, states wrestling with strategic considerations regarding urban warfare must realize that urban warfare is, and will always be, attritional. Moreover, given the rise in urban warfare in modern war, and the likelihood of that trend continuing for the foreseeable future, Western policymakers and military professions would be better served not advocating for smaller, lighter land forces. Rather, they would do right by their respective militaries by investing in robust land forces with increasingly larger margins for error to account for the known, and unknown, challenges associated with rigorous urban warfare.

6 Sieges and the Consequences of the Urbanization of Warfare

Conflict realism finds that sieges are an important part of contemporary war and warfare, and because of their causality in modern war, they will likely continue to loom ominously in future wars. JFC Fuller acknowledges the importance of sieges in war by discussing placing several sieges in military history, to include sieges during the Crimean War and World War I, into their rightful contextual place. Fuller states that a besieger seeks the surrender of a besieged actor, and that treachery, starvation, and assault are the mechanism by which surrender is brought about.[1] The besieger creates that state through either an attack on the defender's moral, an attack on the defender's resources, or an attack on the defender's defenses.[2] But following Fuller, the line goes cold for several decades. Research and writing on sieges withers on the vine.

Jurgen Brauer and Hubert Van Tuyll write that despite occupying a central position in the historical conduct of war, scholars and military professionals have largely ignored sieges.[3] They hypothesize that the community of interest has ignored sieges because they are not glamorous nor fast paced. Sieges are methodical, costly, and undecisive. But, as Brauer and Van Tuyll assert, they are indeed inevitable.[4] The siege's inevitability is examined in detail later in this chapter. However, before arriving at the siege's inevitability, additional context is needed before proceeding to that point.

Nonetheless, implication analysis finds that states and non-state actors have used sieges in nearly every conflict since the end of the Cold War. The siege of Sarajevo, which came right on the Cold War's heels, welcomed the concept into contemporary war, whereas Russia's 2022 siege of Mariupol signals that sieges remain a valid warfighting technique today.

---
1 JFC Fuller, *The Reformation in War* (London: Hutchinson and Company, 1923), 104.
2 Fuller, Reformation in War, 104.
3 Jurgen Brauer and Hubert Van Tuyll, *Castles, Battles, and Bombs: How Economics Explains Military History* (Chicago: University of Chicago Press, 2008), 63.
4 Brauer and Van Tuyll, *Castles, Battles, and Bombs*, 63.

Combatant	Victory	Victory Percentage
Aggresor	36	60%
Defender	18	30%
Multiple	1	1.7%
Ceasefire	1	1.7%
None (ongoing)	4	6.7%

Table 6.1 Victors in Post–Cold War Sieges

The problem with siege is that it elicits a strong negative cognitive bias. On the utterance of the word siege, the audience often mentally drift towards a bygone era of war in which the trebuchet, ballista, and siege engines were the currency of warfare. This anchor bias results in many Futurists and Institutionalist thinkers discrediting the siege's potential position in future war and ignoring the historical and quantitative evidence to indicate the siege's central place in war. The chapter examines sieges and attempts to help illustrate that they are not gone, have not gone away, nor will they be absent in the future of war.

The Logic of Sieges

Sieges recur in modern wars because they are effective and because they are often a response to an emerging situation of power asymmetry between two combatants. We will address the effectiveness factor first.

A historical analysis indicates that sieges develop incidentally. An evaluation of post-Cold War sieges indicates that sieges work. For instance, of the 60 sieges in the post-Cold War period, the aggressor won 60 percent of the time (see Table 6.1). Hence, it follows that combatants use sieges because the siege remains a viable patch to battlefield success.

This finding is not entirely surprising, nor ahistoric. Trevor Dupuy wrote that between 1600 and 1973 the attacker in battle won 61 percent of the time.[5] Today's data on sieges, from a different era of conflict than that queried by Dupuy, roughly match his long study of wins and losses in war.

Nevertheless, sieges are a multifaceted tool for combatants in war. Sieges are typically associated with an aggressor encircling a

---

5 Trevor Dupuy, *Attrition: Forecasting Battle Casualties and Equipment Losses in Modern War* (Fairfax, VA: HERO Books, 1990), 99.

defender. The defender is often a military force, but by virtue of that force being entrapped in an urban area, noncombatants come under siege as well.[6] History shows that the aggressor in a siege is not limited to a state military force. Rather, the aggressor can be any combatant — state, non-state combatant, proxy force, mercenary group, or any other actor with the strength to snare an adversary in a city or town.[7]

Despite International Humanitarian Law (IHL) not defining siege or encirclement, some commenters have pointed to an encirclement as being the required condition making a siege more than just an enveloping attack.[8] Encirclement, however, is not a siege's causal mechanism. In some cases, an aggressor does not possess sufficient size to entirely encircle an urban area, but still finds value in prosecuting a siege.[9] Further, some cases illustrate that terrain makes a full encirclement impossible but helps fill the condition of completing a siege. The lesson here is that sieges take many forms, and like most anything in warfare, they are a pragmatic solution to an emerging military problem.

If an encirclement is not a siege's causal mechanism, then what is? Put simply, combatants use sieges because they accelerate an adversary toward exhaustion and, as Cathal Nolan writes, exhaustion through attrition is how wars are won and lost.[10] Exhaustion is the materiel inability and/or cognitive unwillingness of a combatant to continue fighting. Sieges, unlike head-to-head battles, are one of the best ways a combatant can effectively impose materiel loss on their adversary, while parrying a reciprocal loss.

A siege's effectiveness is the product of the aggressor's ability to exhaust the defender, while not exhausting themselves in the process. As already noted, exhaustion is the currency of a siege – the thing that makes or breaks a combatant's situation during a siege. Similar to Carl von Clausewitz concept of the culmination point, exhaustion is the state in which a combatant can no longer continue to remain physically involved in a conflict due to materiel considerations, or the point at which

---

6 Amos Fox, "On Sieges," *RUSI Journal* Vol. 166, no. 2 (2021): 5.
7 Amos Fox and Beau Watkins, "A Legal Review of Sieges in Modern War," Journal of Military Studies Vol 11, no. 1 (2022): 2022.
8 "The Protection of the Civilian Population During Sieges: What the Law Says," *International Committee of the Red Cross*, 23 January 2024, accessed 16 June 2024, available at: https://www.icrc.org/en/document/protection-civilian-population-during-sieges-what-law-says; Sean Watts, "Siege Law," *Articles of War*, 4 March 2022, accessed 16 June 2024, available at: https://lieber.westpoint.edu/siege-law/.
9 Anthony King, Urban Warfare in the 21st Century (Cambridge: Polity Books, 2021): 27-28.
10 Nolan, *The Allure of Battle*, 588.

their resolve cannot continue to carry the burden of war. An aggressor uses a siege to exhaust a trapped defender, while the defender operates accordingly to stave off exhaustion. Exhaustion is the variable that makes sieges more than just another tactical consideration, but places it, more often than not, in the operational realm.

Why is exhaustion center stage in a siege? Trevor Dupuy's analysis of attrition is useful in this situation. Dupuy writes:

> The faster a front line moves, the lower the casualty rate for both sides...troops advancing rapidly have less time to use their weapons than do troops advancing slowly. When the rate of advance is rapid, more of the soldier's time is spent on the movement itself, and less time is available to bring fire to bear on targets. At the same time, it is difficult to acquire targets during rapid movement, so the defenders are hit less often.[11]

Taking Dupuy's argument in the inverse, one can therefore infer that the opposite holds true for relatively static combat, like that of sieges. The slower a front line moves, as in sieges, the higher the casualty rate.

Resources, time, operations, adversarial action, and international response are the variables upon which exhaustion pivots. An aggressor attempts to deny the defender's access to resources and deny the influx of externally provided resources. In most cases, an aggressor accomplishes this by encircling or blockading the adversary. In concert with resource management, the aggressor strives to deplete the adversary's resources through time elongation, or extending the length of the conflict, while still imposing punishment and resource consumptions. The following heuristic provides one way to think about resource consumption in an adversarial environment:

> Resource expenditure in an adversarial environment (Rx) is equal to quantity of one's force (Qf) plus one's frontage (Ft) plus the number of points of enemy contact along that front (Pc) plus the duration of enemy contact (Dr) divided by one's on-hand resources (Re) plus a combatant's ability to replenish those resources; or (Rp):
> $Rx = (Qf + Ft + Pc + Dr) \div (Re + Rp)$.

---

[11] Dupuy, *Attrition: Forecasting Battle Casualties and Equipment Losses in Modern War*, 101.

This assumption and equation should not be viewed as an irrefutable law, but instead a model to assist in thinking and understanding resource expenditure. Further, friction or entropy can be combatants into the equation to account for the natural tendency of things to not go according to plan.[12]

Considering the importance of exhaustion to military victory, it is therefore important to appreciate that a siege is not just a tactical consideration. Sieges can occur in tactical battles, but in most cases, sieges are more often operational and strategic military activities. Time is a useful independent variable for categorizing sieges into tactical, operation, and strategic consideration. Time is a good metric because it reflects the tension between resource consumption, resupply, operational and strategic resources, and resource distribution across all levels of military activity. Put another way, time is a useful metric because it closely links military activities with exhaustion.

A simple heuristic helps illustrate a siege's inherent logic. If, v = victory, Ar = aggressor's resources, At = aggressor's available time, Ao = aggressor's operation(s), Dr = defender's resources, Dt = defender's available time, Do = defender's operation(s), Ir = international community's response, then,

Additional Reasons to Conduct a Siege

Taking a siege's causality and inherent logic a step further, we can then appreciate that aggressors can also conduct porous sieges. Operation Inherent Resolve's siege of Mosul (2016–17) is an example of this situation.[13] The Iraqi Security Forces (ISF), bolstered by the US-led coalition to defeat the Islamic State, did not possess sufficient force to completely encircle the city of Mosul. The ISF and the US-led coalition nonetheless bifurcated the city and conducted micro-sieges to squeeze the life out of the Islamic State. Anthony King describes micro-sieges as siege operations conducted on a small portion of a larger city. King states that micro-sieges often result from a city's asymmetric size advantage relative to the size of armed force attempting to conduct the siege.[14] That is, today's cities are often too big to be fully encircled, especially when compared to the diminishing size

---

12 Amos Fox, "Getting Multidomain Operations Right: Two Critical Flaws in the US Army's Multidomain Operations Concept," *Association of the United States Army*, Land Warfare Paper 133 (2020): 10.
13 Rick Gladstone, "UN Says Islamic State Executed Hundreds During Siege of Mosul," *New York Times*, 2 November 2017, accessed 26 October 2023, available at: https://www.nytimes.com/2017/11/02/world/middleeast/mosul-atrocities-islamic-state.html.
14 King, Urban Warfare, 28.

of today's military forces. As a result, an armed force must siege selective parts of a city.[15]

An aggressor can conduct a porous siege – like Mosul – for a host of reasons, but there are three primary reasons for conducting a porous siege. First, a porous siege provides the aggressor with intelligence. An open artery to and from a siege affords an aggressor the opportunity to observe what goes into and comes out of a besieged area. That information, in turn, allows an aggressor the opportunity to better understand the defender's capabilities, resource capacity, and force disposition.

Second, an aggressor might use a siege as a punitive action. Leaving a valve open allows the aggressor to keep metered amounts of supply flowing into a besieged area. This allows the defenders, and the affected population, to remain in the fight longer than they would if contact with the outside world was cut entirely. In doing so, the aggressor can actually increase the severity of punishment on the besieged combatant by applying a steady degree of death and destruction on the defender over time. The sieges at Russian sieges of Donetsk Airport (2014–15) and Mariupol (2022), and the Israeli siege of Gaza (2023-present) are examples of this dynamic.[16] In the Russia-Ukraine case, had the Russian military and its proxy forces crimped the siege perimeters entirely, Ukraine's forces would have likely faltered much sooner. But having left a small artery open, the Russian military and its proxy forces inflicted greater punishment by extending the time in which they made Kyiv's forces defend themselves and spend resources to sustain their force. A similar situation befell Hamas in Gaza as the Israel Defense Force (IDF) conducted their sieges in a similar manner.

Third, a porous siege allows a cynical aggressor to provide a nod of nominal good will to the international community and the provisions of IHL. Yet, similar to the first option, this also provides an aggressor the opportunity to gain information on the status of the defender's forces, the overall situation, and force locations.

Comprehending that a combatant engaged in a siege, regardless of their status as the aggressor or the defender, operates on an internal

---

15 King, Urban Warfare, 103-104.
16 Fox, "The Donbas in Flames," 10; Michael Schwirtz, "Last Stand at Azovstal: Inside the Siege That Shaped the Ukraine War," *New York Times*, 24 July 2022, accessed 26 October 2023, available at: https://www.nytimes.com/2022/07/24/world/europe/ukraine-war-mariupol-azovstal.html' "The Latest, The UN General Assembly Votes to Give Palestine More Rights by a Wide Margin," *Associated Press*, 10 May 2024, accessed 15 June 2024, available at: https://apnews.com/article/israel-iran-hamas-latest-05-10-2024-f723eacc54eb85d223e4fa856b9442d0.

Sieges and the Consequences of the Urbanization of Warfare 131

logic of resource maximization at the expense of an adversary's resource depravation, illustrates how sieges can generate operational and strategic results. The Russo-Ukrainian War's siege of Bakhmut, for instance, witnessed Russia dig deep into manpower reserves. The Kremlin authorized its contractual proxy force, Wagner Group, to recruit prisoners into its ranks.[17] This short-term fix allowed Russia to quickly overcome manpower depravation by generating approximately 40,000 combatants in just a few weeks.[18] In Western military terms, Wagner Group's prison recruitment program increased the size of the Russian land forces operating in Ukraine by more than a corps, or by almost four infantry divisions. Although this is but one example, this dynamic illustrates how sieges can (and do) have operational and strategic level consequences in war.

Contemporary Research on Sieges in War

This books examination of sieges begins with the post-Cold War's Yugoslav Wars in the early 1990's and continues through to today's conflicts in Mali, Sudan, Ukraine, and Gaza. In examining post-Cold War sieges, the research focused on answering three simple questions. First, does the aggressor or defender win most often in a siege? Two, considering the combatants, which actors win most often in sieges? Third, how does time factor into the wins and losses of a siege?

Does the Aggressor or Defender Win Most Often in a Siege?

An examination of post-Cold War conflict finds that 60 sieges occurred between the Yugoslav Wars and conflict through June 2024. The aggressor during this period won 36 of the 60 sieges, or 60 percent of the time. The defender, on the other hand, came out on top on 18 times, or 30 percent. The remaining 10 percent of wins are spread across multiple-party victors, ceasefires, stalemates, and ongoing sieges (see Figure 6.1).

The siege of Vukovar (25 August-18 November 1991) is one data point of many that illustrates the aggressor's success rate in post-Cold War

---

17 Christiaan Triebert, "Video Reviews How Russian Mercenaries Recruit Inmates for Ukraine War," *New York Times*, 16 September 2022, accessed 26 October 2023, available at: https://www.nytimes.com/2022/09/16/world/europe/russia-wagner-ukraine-video.html.
18 "Wagner Group Prison Recruits Back in Russia from Ukraine Front Lines Accused of Murder and Sexual Assault," *CBS News*, 27 June 2023, accessed 26 October 2023, available at: https://www.cbsnews.com/news/wagner-group-russia-convicts-prison-recruits-crimes-murder-sexual-assault/.

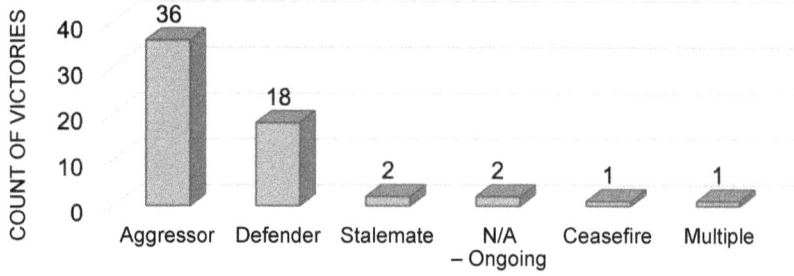

Figure 6.1 Victories by Type of Action

sieges. The siege of Vukovar emerged as an early battle during the Yugoslav Wars and Croatian War of Independence (1 March 1991 – 12 November 1995).[19] Early in the 3-month siege, the Croatian National Guard and the citizens of Vukovar were surrounded by the Yugoslav National Army (JNA) and other Serbian forces. The ensuing righting between the JNA and Croatian National Guard razed the city of Vukovar.[20] After 87 days of continuous isolation and bombardment, the Croatian National Guard and civilian population gave way to the JNA's offensive overmatch.[21] The siege, in hindsight, was a harbinger of things to come in the Croatian War of Independence, which witness tens of thousands dead and thousands more displaced.[22] Furthermore, despite being largely forgotten today, the siege of Vukovar was a seminal event during the Yugoslav Wars because it set the pace for how post-Cold War sieges would unfold over the next 30-plus years.[23]

On the opposite end of time from Vukovar, the Russo-Ukrainian War's siege of Mariupol provides a contemporary data point regarding the aggressor's 60 percent win rate. In Mariupol, Russian forces (the aggressor) were able to isolate, exhaust, and defeat the Ukrainian forces

---

19 "United Nations: International Criminal Tribunal for the Former Yugoslavia," *United Nations*, accessed 10 October 2023, available at: https://www.icty.org/en/about/what-former-yugoslavia/conflicts.
20 Dražen Živić and Iva Šušić Degmečić, "The Battle of Vukovar: A Turning Point in the Croatian "Homeland War,"" *Testimony Between History and Memory* no. 124 (2016): 183-186.
21 Mario Sebetovsky, "The Battle of Vukovar: The Battle That Saved Croatia," Masters Thesis, US Marine Corps Command and Staff College, Marine Corps University: 14-31.
22 "The Breakup of Yugoslavia, 1990-1992," *US Department of State*, accessed 10 October 2023, available at: https://history.state.gov/milestones/1989-1992/breakup-yugoslavia.
23 James Horncastle, "The Death of a City: The Yugoslav Peoples Army Siege of Vukovar, 1991, Refugee Crisis, and Its Aftermath," in Tim Keogh ed. *War and the City: The Urban Context of Conflict and Mass Destruction* (Paderborn, Germany: Brill | Schöningh, 2020), 85-86.

(the defender) after nearly three months of grueling combat in sub-human conditions.[24]

Vukovar and Mariupol are but two of 36 potential examples to illustrate how aggressors are commonly victorious in sieges. This data point is important because it can help policymakers and military leaders make decisions about how to confront problems in future conflicts and wars. At the same time, this information can further help scholars, military analysts, nongovernment organizations, and international organizations who are seeking to better understand the causes of war, how wars are conducted, how wars conclude, and the potential costs associated with real-time events in war.

Considering the Combatants, Which Actors Win Most Often in Sieges?

Having established that aggressors are most often the winners in post-Cold War sieges, it is important to also understand which actors come out on top most often in sieges. This question reframes the initial question about who wins and loses sieges. In doing so, the new question slices wins and losses along the lines of the type of actor and moves away from analyzing wins and losses along attacker-defender lines.

State actors, or state militaries, are win most often in post-Cold War sieges. Of the 60 sieges identified during this period, states won 29 of those occurrences, or 48 percent of the time. What may come as a surprise, however, is that principal-proxy dyads came in second place, winning 14 out of 60 sieges (23 percent). Non-proxy, non-state actors, won 11 of 60 sieges (18 percent). The remaining 11 percent is split between ceasefires, multiple victors, stalemates, and ongoing sieges. See Figure 6.2.

The Syrian Civil War provides an illustration here. During the siege of Homs (6 May 2011 – 9 May 2014), the Syrian army coopted Hezbollah and Iranian irregulars as proxies against the fight against Free Syrian Army forces.[25] Furthermore, as the *Institute for the Study of War* highlights, the Syrian army lacked proper size force to fully encircle the city, and instead

---

24 Christopher Lawrence, *The Battle of Kyiv: The Fight for Ukraine's Capital* (Barnsley, England: Pen and Sword Books, 2023), 153-154.
25 Aron Lund, "What Would the Fall of Homs Mean," Diwan, 24 April 2014, accessed 10 October 2023, available at: https://carnegieendowment.org/middle-east/diwan?lang=en&center=middle-east.

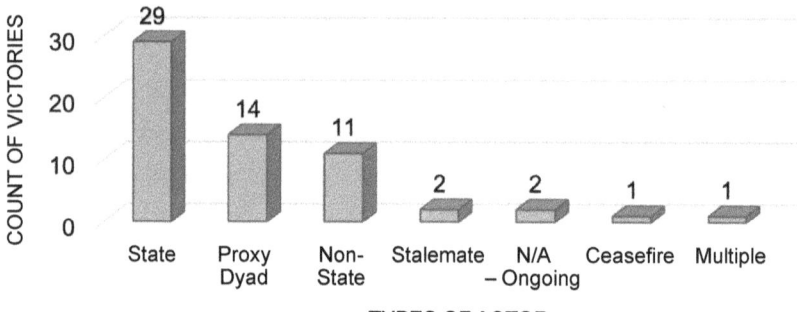

Figure 6.2 Victories by Type of Actor

opted for a porous siege.[26] Nonetheless, the Syrian principal-proxy dyad eventually prevailed in a 3 year long siege. The connection between the principal-proxy day's porous border and the siege's duration are important to understand too. Porousness can allow for more flow of goods in and out of a siege, increasing the durability of the military forces and civilian population within a siege. By a defender increasing their durability, the potential for the siege to elongate – or to grow longer in time, space, and destructive capacity – increases. More on that momentarily.

Time and Wins and Losses in Sieges

We will address time and wins and losses in sieges using a set of variables. First, we will use the type of action and type of actor as our dependent variables for this assessment. Time will serve as our independent variable. Time will be broken into four categories: (1) 1 day or 1 month, (2) 1-6 months, (3) 6-12 months, and (4) 12 months or longer. Second, we will use those dependent variables and the independent variable to examine what type of action (that is, offensive or defensive operations) our 60 post-Cold War sieges favored over time. Third, we will examine what type of actor (state, principal-proxy dyad, non-state actor, et cetera) our 60 sieges favor over the same time period. Fourth, we will compare the actor and action data and make inferences based on the comparison.

---

[26] "The Syrian Army Renews Offensive in Homs," *Institute for the Study of War*, 5 July 2013, accessed 10 October 2023, available at: https://www.understandingwar.org/backgrounder/syrian-army-renews-offensive-homs.

Wins by action, over time

Before examining the data, a few points of clarification are required. Within the action category, six categories exist. The first is aggressor. The aggressor is the actor, regardless of their status as a state military, a non-state actor, a principal-proxy dyad, or any other taxonomy, who is conducting the siege. Besieger is another term appropriate for aggressor. Conversely, the defender is the actor under siege. Defenders can equally be referred to as the besieged, or as being besieged. Those two categories are the most prevalent when examining actions in siege operations. Nonetheless, ceasefires, stalemates, and in a few cases ongoing sieges are additional actions to account for. In one case – the siege of Aleppo (19 July 2012 – 22 December 2016) from the Syrian Civil War (15 March 2011 – present) – multiple actions occur, so that is accounted for with the phrase 'multiple'. These terms – aggressor, defender, ceasefire, stalemate, multiple, and ongoing – are how actions are articulated within this section.

Ten sieges – or 17 percent of all 60 post-Cold War sieges – occur within the 0-1 month range. Within this time frame, the aggressor prevailed 90 percent of the time and the defender only 10 percent. 21 sieges, or 35 percent of the 60 post-Cold War sieges, occurred During the 1-6 month time frame. The aggressor one 12 of those 21 instances of siege, whereas the defender came out on top in 7 engagements. The other 2 sieges are split between an ongoing siege at Timbuktu as part of the Mali War (17 January 2012 – present) and a ceasefire at Dammaj from the Yemen Civil War.[27]

We can therefore see that the aggressor wins most often in sieges. However, if a defender can stretch a siege out to the 6-12 month mark, their odds of winning significantly increase. Yet, the window is small because after the 12 month mark, the aggressors' chances of success are as great at any point during a siege (see Figure 6.3). Nonetheless, wins by action across time are only one way to examine the evidence of post-Cold War sieges. Looking at the information through the lens of wins by actor across time is also helpful.

Wins by actor, over time

As noted previously, states win 50 percent more often in sieges (29 wins) than proxy dyads (14 wins), who take second place. But, when using the same time taxonomy used in in the previous section, examining the data

---

27 Post-Cold War Data Set_26 October 2023.

136 Conflict Realism

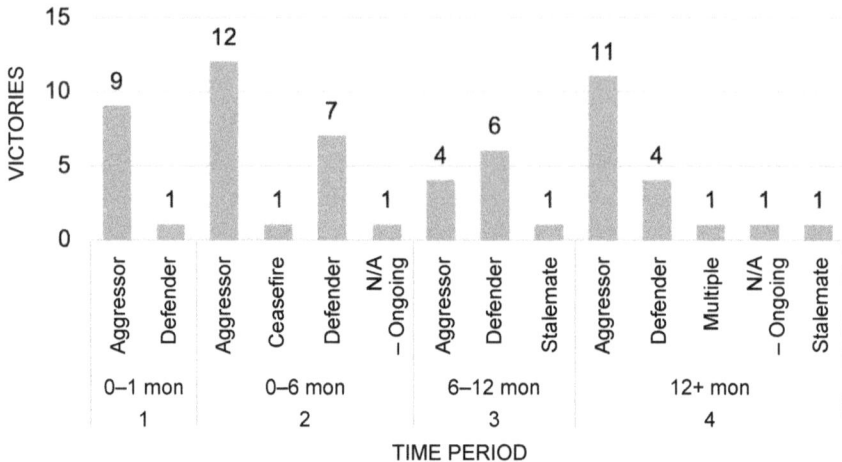

Figure 6.3 Victories across Time, by Action

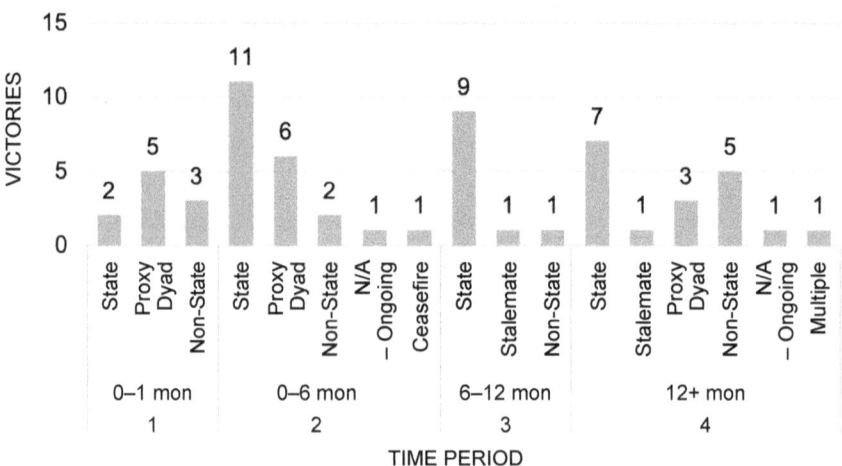

Figure 6.4 Victories by Actor, over Time

across time, the wins by actor yield interesting results. Figure 6.4 is used to illustrate the data for the section.

Looking at the same 10 sieges during the 0-1 month time period through the lens of actor finds that state actors won only 20 percent of the time, whereas proxy dyads were more profitable, winning 50 percent of the sieges. Non-state actors also tallied a higher winning percentage, coming in with 30 percent of the grand total.

During the 1-6 month period, the data represents what many onlookers might expect to see. During that period, we recorded 21 sieges. Of those 21 sieges, state actors won 11 times (53 percent). Proxy dyads came in second with 6 wins (29 percent), followed by non-state actors, an on-going siege, and a ceasefire making up the remaining 20 percent.

The time findings hold true during the 6-12 month period – states are dominate during this period. In fact, the state is most dominate during this time frame. Of the 11 sieges lasting between 6-12 months, states won 9 times (82 percent). The remaining two wins are evenly distributed between a non-state actor and a stalemate. A point of emphasis here. For states looking to effectively use a siege, it is perhaps helpful to try and hit this relative sweet spot. When compared to the other three time windows, a state's chance of victory is almost 30 percent higher in this time period than at any other time. Thus, thoughtful states should attempt to situate siege operations into this 6-12 month sweet spot if they want to operate according to good historical trends. Moreover, any actor attempting to overcome a state engaged in a siege, but actively seeks to either bring the siege to a close before 6 months or after a year, at which time the state's chances of success drop significantly.

The final time frame, 12-plus months, aligns with what one might expect. 18 sieges fit within the time frame. States tallied the most wins (7 wins; 39 percent), while non-state actors come in second (5 wins; 28 percent). Proxy dyads, which have performed quite well throughout the sieges, finish third (3 wins; 17 percent). The remaining 28 percent is distributed evenly amongst stalemate, ongoing sieges, and multiple actor wins.

Data clearly illustrates that states dominate across the middle of the siege timeline. To be sure, 32 sieges occur between 1-12 months. States won 20, or 63 percent, of those sieges. Thus far the data collected has not been able to explicitly explain why states thrive in the middle of the siege timeline, nor why they appear to struggle on the short and long ends of the spectrum.

Nevertheless, one can assume that the middle band — between 1 and 12 months — better suits a state's logistics network. At the short end of less than a month, states might not perform as well because they are quickly overcome by surprise. On the long end of the spectrum, or longer than a year, a state's power might wain due to flagging political or domestic support. Nonetheless, further research is required to provide better educated insights into that problem.

Defining the Character of a Siege

Sieges occur against a land force in relatively open terrain or against a land force in closed (urban) terrain. In most circumstances, sieges are a retort to a situation's variables. A siege in open terrain, for example, is the result of one of four situations. First, it is the result of Combatant A wedging itself between Combatant B and the force to which Combatant B belongs. In cutting off Combatant B from its main body, Combatant A then isolates and encircles Combatant B, placing it in position for the ensuing siege. Second, it is the result of Combatant B moving too far ahead of its adjacent units or main body. This allows Combatant A to fix it in place and encircle it, while fending off Combatant B's adjacent support. Third, it results from Combatant B encountering Combatant A, which is numerically superior to Combatant B, on the battlefield. Combatant A overwhelms and envelops Combatant B, like the arms of an octopus wrapping around its prey. Fourth, Combatant A, which possesses a disproportionate degree of operational or tactical mobility than Combatant B, deftly sweeps around Combatant B, isolating the less mobile combatant in place.

Sieges in open terrain, however, have not been identified in post-Cold War analysis of war. The reason for this is not clear. The lack of sieges in open terrain might be a product of the shrinking size of state armies and small armies using urban areas are their primary base of operations, instead of having large elements of force scattered throughout the countryside.

In the past, when state armies were larger, one army catching an element of an adversary's army in the open, and then encircling it, was relatively common. The German army's operational design and tactics were essentially based on this idea from the time of Frederick II (1712-1786) through the end of World War II (1945). Robert Citino offers that throughout the period between Frederick II and World War II, the German army sought to separate an adversary's force from their support network.[28] In doing so, the German army sought to subsequently encircle their adversary, and then make quick work of the enemy through concentrically annihilating each layer of the defending force.[29] This concentric annihilation continued until the adversary acquiesced or ceased to exist as a fighting force.[30]

---

28 Robert Citino, *The Death of the Wehrmacht: The German Campaigns of 1942* (Lawrence, KS: University of Kansas Press, 2007), 3-4.
29 Citino, The Death of the Wehrmacht, 3-4.
30 Citino, *The Death of the Wehrmacht*, 3-4.

The Germans referred to this technique as the *kesselschlact*. Citino writes that the *kesselschlact*, however, was not a siege. According to Citino, the German army encircled their adversaries merely as, "The preparation to attacking him from widely separated compass points, chopping up the encirclement into smaller portions, and finally smashing him."[31] Citino states that the Germans referred to the post-encirclement operations as concentric, or *konzentrisch*, operations.[32]

Cold War and pre-Cold War logic about the interrelationship between force structure and operations suggest that Citino is correct about the *kesselschlact* not being a siege. Using today's logic that small armies conduct micro-sieges, however, makes the connection between the German's *konzentrisch* operations and sieges quite apparent.

Sieges in closed terrain, on the other hand, often unfold in a differing manner. Closed sieges occur in urban terrain and are often the product of a combatant conducting an urban defense or retrograding from position outside a city into an urban area. As is often the case, the combatant seeking refuge in the urban environment is physically weaker than the aggressor. The weaker combatant maximizes the latent strength of urban terrain to functionally and positionally dislocate the aggressor's strength and advantages through smart tactics and operations. The natural response to this is to encircle and isolate the defender and gradually work to bring about that combatant's acquiescence through brute strength, gradual attrition, and targeted killing.

The closed, or urban siege, takes one of three basic forms. The first form, which can be referred to as the 'Parity' option, entails a defender seeking refuge in an urban area. In this option, the defender is the focal point of analysis. In the Parity option, the defender is often a comparatively weaker force, such as a non-state combatant or a smaller, less capable state military force. The defender's operational and tactical goal is thus to neutralize or at least offset the aggressor's strength, thereby creating situational parity with the aggressor. Strategically speaking, the defender's goal is to stave off an existential crisis and continue to exist.

The Second Chechen War's Battle of Grozny (25 December 1999 – 6 February 2000) exemplifies this situation. Having learned from its missteps during the First Chechen War and its battle for Grozny, Russian

---

31 Robert Citino, *The Wehrmacht's Last Stand: The German Campaigns of 1944-1945* (Lawrence, KS: University of Kansas Press, 2017), 20.
32 Citino, *The Wehrmacht's Last Stand*, 21.

military forces used new tactics during the Second Chechen War's fight for Grozny. Russian forces, for instance, utilized a siege to bend Chechen will – through resource exhaustion – during the fight for Grozny. Destruction was the central feature of Russia's siege. Russian forces encircled Grozny and slowed choked it into submission by cutting off resources and supplies from external areas. Concurrently, Russian forces sought out and eradicated Grozny's Chechen defenders. The siege, an incredibly destructive endeavor, was highly influential on the war's outcome.[33]

Second, urban sieges are the result of an attempt to neutralize urban insurgencies or urban partisan movements. In this situation an insurgency or partisan movement operates in the city and amongst its people to generate and maintain support. The targeted state combatant, or combatant aligned against the insurgency or partisans, employs a siege to bring the movement to heel. The sieges of Fallujah (4 April – 1 May 2004 and 7 November – 23 December 2004) during Operation *Iraqi Freedom* are examples of this form of urban siege.[34]

The third form of urban siege is a response to an adversary's urban defense. This can happen among state combatants, non-state combatants or any combination therein. In this situation one combatant defends from an urban area, whether intentionally or by getting caught in urban terrain before being able to move beyond that area. The aggressor, assessing that they are in a dominant position, encircles the isolated defender and begins the siege. History is littered with many examples of this siege. The two-day siege of the US Embassy in Baghdad (31 December 2019 – 1 January 2020) is an instructive tool to spotlight this form's non-standard method. In this siege, Iranian-aligned militia groups, to include Kata'ib Hizbullah, encircled the embassy and held its occupant's hostage.[35] The siege was the result of escalating tension between the US and Iraq throughout December 2019. The siege withered because the US inserted a company of Marines into the embassy complex, dispatched attack aviation to overwatch the

---

33 Olga Oliker, *Russia's Chechen Wars 1994-2000: Lessons from Urban Combat* (Monterey, CA: RAND, 2001), pp. 22–28.
34 Jonathon Hayes, *Combat Stories Map: A Historical Repository and After Action Tool for Capturing Stories, Storing, and Analyzing Georeferenced Individual Combat Narratives* (Monterey, CA: Naval Postgraduate School, 2016), 1-7, https://apps.dtic.mil/dtic/tr/fulltext/u2/1026636.pdf, accessed April 15, 2021.
35 Michael Gordon, Isabel Coles and Ghassan Adnan, 'Trump Vows Retaliation After Iranian-Backed Militia Supporters Try to Storm US Embassy in Baghdad', *Wall Street Journal*, 31 December 2019.

compound, and deployed another 750 soldiers from the US to Iraq.[36] The Berlin Blockade and the Western Allies' response is another example of this type of strategic siege.[37]

Clarifying Misconceptions About Sieges

In conducting the research to build the siege dataset for this chapter, a handful of misconceptions kept emerging as we consulted with scholars, analysts, military practitioners, and consulted with the limited scholarship on the subject. Often ignored altogether in academia, bias and wrongminded ideas cloud the understanding of sieges in public and practitioner discourse. Thus, it is important to discuss the misconceptions in the hopes of preventing their continued use and abuse in relation to studying and understanding both war and warfare.

Misconception 1: Sieges and siege warfare are synonyms terms

A siege is a tactic or operation (which can stretch into a campaign) that a combatant uses against an adversary for a variety of reasons. Because sieges can be a tactic, an operation, or a campaign, they are not just tactical considerations. To be sure, sieges on the operation and campaign side of the equation are operational level military activities, which can fuel strategic results and accelerate changes in the conduct of a battle, operation, or campaign. On occasion, sieges can also create decisive effects – that is, an outcome so significant that it: 1) causes a combatant to significantly alter their military operations, 2) causes a policymaker to change their policy regarding a specific conflict, or 3) a combination of both 1 and 2.

The German success in the Ardennes in December 1944, which caught the Allies completely by surprise, is an example of a decisive military operation. The German offensive resulted in the theater commander, General Dwight Eisenhower, directing a wholesale change in the Allies' operational plan.[38] After a crisis management meeting in Verdun on 19 December 1944, Eisenhower pulled Lieutenant General George Patton's

---

[36] Conor Finnegan et al., 'Pentagon Expects to Deploy Troops to Mideast After Iraqi Protesters Assault US Embassy, Torch Guardhouse to Protest Airstrikes', *ABC News*, 1 January 2020.
[37] Daniel Altman, "Advancing without Attacking: The Strategic Game Around the Use of Force," *Security Studies* (Vol 27, no. 1, August 2018), p. 16-18.
[38] Kevin Hymel, *Patton's War: An American General's Combat Leadership, Volume 2, August-December 1944* (Columbia, MO: University of Missouri Press, 2023), 299-300.

3rd Army out of its southeastward advance, turned it 180 degrees to the north, had it march approximately 100 miles and move directly into a fight at Bastogne to relieve Courtney Hodges' 1st Army and the 101st Airborne Division (an element of the theater reserve), both of which were under siege by the Germans.[39]

On the other hand, the causality between Russia's failed Kerensky Offensive during World War I, and their subsequent disengagement from the conflict is an example of political decisiveness. Russia's ill-fated Kerensky Offensive (July 1917) failed to exploit any of the marginal successes from the preceding Brusilov Offensive (June 1917).[40] The Russian military's failure to facilitate acceptable outcomes and battlefield progress during the Kerensky Offensive precipitated Russia's withdraw from the conflict.[41] The failed offensive was not the only cause for Russia's withdraw, but within the geopolitical situation in which Russia found itself, the operation can be essentially considered to have thrown gasoline on an already expanding fire. Russia's departure from the war was formalized on 3 March 1918 in the Treaty of Brest-Litovsk.[42]

Nevertheless, decisiveness is not a common outcome from sieges. Only five – or 8 percent – of post-Cold War sieges generated strategically and politically decisive results. These sieges include The Bosnian War's siege of Sarajevo, the Second Chechen War's siege of Grozny, the Russo-Ukrainian War's sieges of Ilovaisk and Debal'tseve, and the Philippines' siege of Marawi.[43]

Moreover, modern sieges are not a 'way of war', a 'way of warfare', nor a type of warfare. The phrases 'way of war' and 'way of warfare' implies that a corpus of warfighting, underpinned by a common logos, guides any 'way of…' assertion. To that end, any 'way of warfare' is often bolstered by force structure, weapons, doctrine, and training alignment. Modern sieges are often a pragmatic response to an emerging set of specific conditions within a conflict.

A survey of the doctrine, force design, and education of contemporary Western militaries finds that they are not organized, educated, nor equipped for anything resembling 'siege warfare'. Instead, sieges are generally

---

39 Hymel, *Patton's War*, 304-306.
40 Cathal Nolan, *The Allure of Battle: A History of How Wars Have Been Won and Lost* (Oxford: Oxford University Press, 2017), 388.
41 Nolan, *The Allure of Battle*, 388.
42 Nolan, *The Allure of Battle*, 388.
43 *Post Cold War Siege Dataset*.

not accounted for, but when they are, they are rolled into urban warfare considerations. Further, some analysts place increasingly restrictive caveats on what constitutes a siege. Encirclement is one such caveat. Many commenters suggest that for a situation to be a siege, the aggressor must have completely encircled a defender.[44] Caveats such as this are not necessarily true, nor are they helpful when attempting to better understand sieges.

Having established that sieges are not a way of war, a war of warfare, nor a type of warfare, it is important to place sieges within the temporal and spatial context of war. Sieges can occur within battles, operations, or campaign, or be any of those types of engagement. For a siege to be classified as a battle (siege-battle), it must be a singular event – that is, the siege is the only significant activity occurring within a specific place and time within a conflict. Importantly, for a siege to be a battle it must also be of relatively short duration. For classification purposes, 'short duration' is defined as less than 30 days. This time period is prescribed because it aligns with logistics considerations. 30 days is about the minimum a military force or civilian population can survive with minimal logistics concerns, with a few caveats. For environments in which things like water and food are already sparse, then logistics considerations become dangerous far quicker than they do in places in which access to water and food are far more ample. Moreover, siege-battle configurations often emerge as a pragmatic response to developing battlefield situations.

The Russo-Ukrainian War's siege of Ilovaisk (August 2014) is a good example of this taxonomy. At Ilovaisk, Russian and DPA forces encircled Ilovaisk, which was home to several Ukrainian battalions. The Russian and DPA forces cut the city's power supply and other utilities early in the battle. They then pummeled the city and the Ukrainian forces with vast amounts of cannon, rocket, and missile fire, while conducting armored and light infantry attacks against the defending Ukrainian forces.[45] The siege was lifted only when Russian and DPA forces double-crossed Ukrainian force during an offer of safe passage out of the city.

---

44 John Spencer, "The Battle of Marawi," *Urban Warfare Project* (Modern War Institute podcast), 4 March 2021, https://podcasts.apple.com/us/podcast/the-battle-of-marawi/id1490714950?i=1000511672529.
45 Amos Fox, "The Siege of Ilovaisk: Manufactured Insurgencies and Decision in War," *Association of the United States Army*, Land Power Essay 21-2 (2021): 2-3.

Misconception 2: Sieges require full encirclement or isolation of an enemy

Sieges are situationally dependent and therefore predicating a siege on something like complete encirclement or isolation within a geographic pocket does not stand up to the logic of siege operations. Sieges are a multi-faceted tool for combatants in war. Sieges are typically associated with the encirclement of an aggressor against a defender. The defender is historically a military force, but by virtue of that force being entrapped in an urban area, non-combatants come under siege as well. Aggressors can be any combatant – state, non-state combatant, proxy force, mercenary group, or any other combatant with the strength to snare an adversary in a city or town.

Encirclement, however, is not a siege's causal mechanism. In some cases, an aggressor does not possess sufficient size to entirely encircle an urban area, but still finds value in prosecuting a siege. Therefore, they knowingly execute a porous siege. Operation Inherent Resolve's siege of Mosul is an example of this situation. The Iraqi Security Forces (ISF), bolstered by the US-led coalition to defeat the Islamic State, did not possess sufficient force to completely encircle the city of Mosul. The ISF and US-led coalition nonetheless bifurcated the city and conducted micro-sieges to squeeze the life out of the Islamic State.[46] In many other cases, however, adroit combatants will leave a portion of a siege open. The aggressor might do this for a variety of reasons.

First, a porous siege provides the aggressor with intelligence. An open artery to and from a siege affords an aggressor the opportunity to observe what goes into and comes out of a besieged area. That information, in turn, allows an aggressor the opportunity to better understand the defender's capabilities, resource capacity, and force disposition.

Second, an aggressor might use a siege as a punitive action. Consequently, leaving a valve open allows the aggressor to keep metered amounts of supply flow into a besieged area, allowing the defending force and the affected population to remain in the fight longer than they would if contact with the outside world were cut entirely. In doing so, the aggressor can actually increase the severity of punishment on the besieged combatant by applying a steady degree of death and destruction on the defender over time. The siege at Donetsk Airport and Mariupol are examples of

---

[46] Amos Fox, "The Mosul Study Group and the Lessons of the Battle of Mosul," *Association of the United States Army*, Land Warfare Paper 130 (2020): 2-5.

this dynamic. In each case, had Russian and Russian proxy forces closed the perimeters entirely, the Ukrainians would have likely faltered much sooner. But having left a small artery open, Russian and Russian proxy forces inflicted greater punishment by extending the time in which they made Kyiv's forces defend themselves and spend resources to sustain their force.[47]

Fourth, a porous siege allows a cynical aggressor the ability to provide a nod of nominal good will to the international community and the provisions of international humanitarian law. Yet, similar to the first option, this also provides an aggressor the opportunity to gain information on the status of the defender's forces, the overall situation, and force locations.

If an encirclement is not a siege's causal mechanism, then what is? Put simply, combatants use sieges because they accelerate an adversary toward exhaustion and, as Professor Cathal Nolan writes, exhaustion through attrition are how wars are won and lost.[48] Exhaustion is the materiel inability and/or cognitive unwillingness of a combatant to continue fighting. Sieges, unlike head-to-head battles, are one of the best ways a combatant can effectively impose materiel loss on their adversary, while parrying the reciprocal loss.

Considering the previous point, it is important to note that a siege is not just a tactical consideration. Sieges can occur in battles, but in most cases, sieges are more often operational and strategic military activities. Time is a useful independent variable for categorizing sieges into tactical, operation, and strategic consideration. Time is a good metric because it reflects the tension between resource consumption, resupply, operational and strategic resources, and resource distribution across all levels of military activity. Put another way, time is a useful metric because it closely links military activities with exhaustion.

Misconception 3: Sieges require starvation

Though not clearly articulated in literature today, many commentators and conference attendees suggest that starvation must be employed as part of

---

47 Amos Fox, "'Cyborgs at Little Stalingrad': A Brief History of the Battles of Donetsk Airport, 26 May 2014 to 21 January 2015," *Institute of Land Warfare*, Land Warfare Paper 125 (2019): 5-11; Michael Schwirtz, "Last Stand at Azovstal: Inside the Siege That Shaped the Ukraine War," *New York Times*, 24 July 2022, accessed 24 October 2023, available at: https://www.nytimes.com/2022/07/24/world/europe/ukraine-war-mariupol-azovstal.html.
48 Nolan, *The Allure of Battle*, 588.

a general encirclement. While starvation of military forces is admissible under IHL, it is prohibited by Rule 53 of customary IHL. Rule 53 states in international armed conflict (i.e., war between states) it is lawful to starve hostile belligerent forces, but provisions to protect and care for civilian populations are a must.[49]

Militaries can thus conduct a siege without starving the civilian population, while attempting to starve their military combatant. Likewise, a military cannot starve the population and not starve their military combatant, and still conduct a siege. Therefore, using starvation as a variable to prove or disprove a siege is incorrect.

Misconception 4: The effects of a siege are felt in one place

It is quite easy to fall into the trap of assuming that a siege only impacts the area under siege, however, post-Cold War siege analysis clearly affirms a contrary position. Analysis illustrates that basing, support, and firing locations are prone to a relatively reciprocal level of destruction as the besieged location.[50] Sieges often occur in urban areas because of the protection from observation an urban area provides, in combination with most combatants' unwillingness to inflict intentional civilian casualties and collateral damage. Nonetheless, Henry Langston states that the cities, towns, and villages that are adjacent to areas under siege are often eviscerated in the process.[51]

From a purely theoretical standpoint, the destruction of towns, cities, and villages adjacent to areas under siege makes sense. A besieged combatant will certainly counter-strike adversarial firing locations if they are able to get a bead on the location from which aggressor's strikes originate. Moreover, if an aggressor's basing and support locations are known to the besieged combatant, the besieged forces will also likely strike at those locations. The goal of the besieged actor's strikes at firing, support, and basing locations is to disrupt the aggressor's operations, and perhaps even nudge the aggressor towards exhaustion. Considering that the aggressor's chances of winning a siege drop exponentially as time

---

49 "Rule 53, Starvation as a Method of Warfare, Customary International Humanitarian Law," *International Humanitarian Law Database*, accessed 24 October 2023, available at: https://ihl-databases.icrc.org/en/customary-ihl/v1/rule53
50 Fox, "The Siege of Ilovaisk," 1-2.
51 Henry Langston, "Reflects on Russia's 2014-2015 Donbas Campaign," *Revolution in Military Affairs* (podcast), 4 March 2024, https://shows.acast.com/650105b75a8d440011ecd53c.

moves beyond a month, while the besieged combatant's odds historically improve to almost a 60 percent success rate by a year, a defender benefits from dragging the conflict out by disrupting the aggressor's operations and eroding their materiel and human resources.

In Ukraine, for example, the sieges of the Donbas Campaign are telling. During the siege of Ilovaisk, one of the war's first sieges, the towns of Hrabske, Mnohapillia, and Pokravka were pulverized. Between being on the receiving end of shelling and suffering the effects of soldiers using the cities resources for protection and to live off the land, those towns were one aspect of the siege of Ilovaisk's collateral damage.[52] The battle of Donetsk Airport, which began right on the heels of the battle of Ilovaisk, experienced a similar situation. The small village of Pisky, located a mile or so northwest of the airport, was all but wiped from the map during the siege of the airport.[53]

It is therefore important that when thinking about the conduct, impact, and long-term effects of a siege, policymakers, military practitioners, and analysts must seriously consider how operations inside the siege will impact the situation outside of a siege. Failure to do so can lead to serious consequences for gaining and/or maintaining support from the location population. The local population, as the wars in Afghanistan, Iraq, Syria, and Ukraine illustrate, are one of the most important variables in warfare.

Synthesis: Characteristics of Modern Sieges

Having examined the primary misconceptions concerning modern sieges, it is now useful to discuss what modern sieges are. Research finds that a siege's characteristics can be grouped along six basic lines. This section briefly examines each characteristic.

Characteristic 1: Modern sieges are attritional battles of encirclement and position

In contemporary sieges, encirclement and position are the conduits of destruction-based combat which causes a siege to generally develop into a battle (or operation) of attrition. Combatant A's goal of a siege is to force Combatant B to submit to Combatant A's political-military intentions.

---

52 Amos Fox, "The Siege of Ilovaisk," 2-3.
53 Amos Fox, "'The Cyborgs at Little Stalingrad': A Brief History of the Battles of Donetsk Airport," *Association of the United States Army*, Land Warfare Paper 125 (2019): 3-4.

Destruction is the currency of a siege This is because destruction-based warfare, unlike softer approaches, instills tangible fear. Moreover, encirclement seeks to deprive the besieged combatant of the means to revitalize their forces by denying their ability to sufficiently supply themselves.

Further, adroit combatants use perilous terrain to help encircle an enemy combatant and create an advantageous situation for the aggressor. Cities are the most common place this situation occurs. When a weaker force contracts into the perceived safety of a city, many stronger actors attempt to turn the situation to their own advantage by encircling the city and cutting the lines of support into and out of the city. They do this to deprive the defending combatant of the ability to resupply their forces, evacuate casualties, and gain and maintain situational awareness regarding the tactical and operational situation.

These variables – destruction, encirclement, and position – must be viewed in conjunction with one another because they form the casual mechanism that unlocks any siege's potential.[54] To that end, sieges seek to destroy an enemy combatant and their support network until a) the enemy combatant's force acquiesces, b) the combatant's military force is unable to continue resisting, or c) the enemy combatant's political leadership intervenes.[55]

Characteristic 2: Modern sieges are often a response to a situation's variables

The dataset that supports most of the information in this chapter indicates that most post-Cold War era sieges are often an emergent response to an evolving military situation. Of the 60 sieges identified during the post-Cold War period, only two appear to have been preplanned events. This supports the assertion that siege warfare – a way of warfighting codified with a set tactics, operations, and forces specifically built for the purpose of conducting sieges – is a relic of a bygone era of war, but sieges are a pragmatic mainstay of war.

Modern sieges, like those at Ilovaisk or Mosul, develop relatively organically to the situation at hand. In August 2014 at the Ukrainian city of Ilovaisk, Ukrainian forces had forced their way into the city, but they

---

54 Fox, On Sieges, 8-9.
55 Fox, On Sieges, 8-9.

possessed forces insufficient to hold open a route into and out of the city.[56] Russian proxy forces realized the Russian forces precarious situation and worked to encircle the city and thus trap the Ukrainian forces therein.[57] The challenge of keeping the siege sealed tightly improved when Russian land forces arrived on the scene in later August. Then both Russian proxy forces and Russian land forces were able to prevent Ukrainian reinforcements and batter the withering Ukrainian units into submission, and pressure Kyiv toward political decisions aligned with Moscow's interests. In effect, the Russian proxy forces and Russian land forces followed the siege playbook – find a combatant in a disadvantaged physical, temporal, and cognitive position, encircle the combatant, deny their ability to reinforce or resupply themselves, and then bludgeoning their forces with indirect and direct fires until they give up, or their government gives up. In the case of Ilovaisk, Ukraine's armed forces surrendered and negotiated a planned withdraw and politically, Kyiv settled for a nominal ceasefire – the Minsk Protocol. Russian proxy forces honored neither the ceasefire nor the planned withdrawal, turning Ukraine's retreat from Ilovaisk into a slaughter.[58]

Mosul followed a similar pattern. Islamic State fighters consolidated in Mosul in a bid to hold onto their capital in Iraq. The Iraqi security forces and the US-led Coalition quickly encircled the city. Not possessing sufficient force to completely encircle Mosul and conduct a systematic clearance of the city – eliminating Islamic State fighters city block by city block – resulted in a porous siege.

A porous siege is one in which the besieging combatant does not fully encircle an enemy combatant. This can result from the besieging combatant possessing insufficient land forces in relation to the objective (if an urban area) or the decision to leave arteries to and from the besieged location open. The decision to not entirely seal off a besieged combatant might be made to draw the conflict out over time and thus inflicting more costly losses on the enemy than they would if they quickly defeated them. Likewise, the decision to not seal off a besieged combatant might also be done to adhere to – or give the appearance of adhering to – IHL.

Porous siege aside, the Islamic State fighters elected to stand their ground in Mosul and attempt to militarily defeat the Iraqi security forces

---

56 Fox, "The Siege of Ilovaisk," 2-3.
57 Fox, "The Siege of Ilovaisk," 2-3.
58 Fox, "The Siege of Ilovaisk," 2-3.

and US-led Coalition.[59] These decisions led to the Iraqi security forces and the US-led Coalition systematically eradicated the Islamic State in the city.[60] As a result, the fighting for control of Mosul was brutal and destructive. After nine months, the city of Mosul and its people were pulverized and yet the Islamic State were defeated in Iraq.[61]

As Ilovaisk and Mosul illustrate, sieges often develop from emergent battlefield conditions. Those conditions include situations in which an enemy combatant's position (or where they are on the battlefield in relation to their adversary) is advantageous to oneself. Further, these conditions include situations in which an enemy combatant is operating at a tempo that favors oneself. For instance, if a military force is operating slow and methodically, a combatant that accelerates the encirclement and concentric destructive of a besieged enemy combatant might be able to tally a quick win. Conversely, a combatant operating at a high tempo might be able to be lured into a trap of terrain, which subsequently contributes to an adroit enemy springing an encirclement operation on the trapped combatant. The siege – often pragmatic responses to evolving battlefield conditions – thus presents a safe, more efficient and cost-effective method to fight in many cases.

Characteristic 3: Modern sieges are a game of dominance and resource expenditure in an adversarial context

Dominance is not a correlation of forces and means comparison. Dominance is the ability to impose one's will against an enemy combatant, or any other actor within a specific geographic area for a specific period of time. Thus, dominance possesses a relational quality – it measures strength between multiple actors. Dominance possesses an adversarial context – it measures the application of strength against another actor, or actors, in pursuit of one's own objectives. Because applying strength requires to application of resources over time, dominance possesses temporal and resource qualities too. In total, dominance can be thought to contain the following variables: resources, time, enemy action, and self-sustainment. Taking this logic a step further, the following equation can be used to help simplify the understanding of dominance. Dominance (D) equals one's

---
[59] Jeff Martini, et al., *Operation Inherent Resolve: US Ground Force Contributions* (Santa Monica, CA: RAND Corporation, 2022), 125-127.
[60] Martini, et al., *Operation Inherent Resolve: US Ground Force Contributions*, 125-127.
[61] Martini, et al., *Operation Inherent Resolve: US Ground Force Contributions*, 125-127.

resources (*Re*) plus time (*Ti*), divided by an enemy's action (*Ea*), plus self-sustainment (*Su*); or $D = (Re+Ti) \div (Ea+Su)$. Understanding this struggle – and its associated equation – helps policymakers, military practitioners, and analysts understand the underlying dynamics at work in sieges as a combatant strives to either impose dominance on another combatant or overcome a dominated situation.

Sieges are equally a game of resource consumption, in an adversarial context, across time. This is alluded to in the discussion on dominance. To be sure, the dominance equation's emphasis on one's resources (*Re*) and self-sustainment (*Su*) are where this dynamic resides. Nonetheless, resource consumption in an adversarial context across time is sufficiently important to sieges to warrant further examination too.

As with the concept of dominance, resource consumption in an adversarial context (*Rx*) consists of several variables. First is the quantify of one's force (*Qf*). This is a relative variable, and it is viewed in relation to the situation at hand. The *Qf* is the size of one's force in relation to military operation being conducted. If an engagement, the *QF* is a smaller unit, whereas if the scale of operations grows to something in which field armies are the organizations being measured, then the *QF* might be quite large.

The *QF* is balanced against the force's assigned or assumed frontage (*Ft*), plus the number of points of enemy contact along the front (*Pc*), plus the duration of the contact with the enemy (*Dr*). Viewed collectively, these variables equal the basic building blocks of a military operation (*Op*). This balance can be qualified with the following equation: $Op = Qf+Ft+Pc+Dr$.

The previous paragraph, however, does not account for how an operation is impacted by resources. Thus, two additional variables are needed to better make this heuristic representative of reality; those variables are on-hand resources (*Re*) and one's ability to replenish their resources (*Rp*). One's on-hand resources are often small and just enough to support existing operations. This variable is relative to the combatant being observed. A small tactical unit – say a company or a battalion – will have a vastly different array of on-hand resources than a corps or a field army. Likewise, the same relativity exists regarding a combatant's ability to replenish their supplies. Small tactical units are dependent on their higher headquarters and attached logistics support, whereas larger military forces possess more internal replenishment mechanisms. Nonetheless, these two variables – on-hand resources (*Re*) and one's ability to replenish their resources (*Rp*) – must be applied to an operation (*Op*) to truly appreciate a force's resource expenditure in an adversarial context. Thus, the following equation can be used to help illustrate this idea: $Rx = Op \div (Re + Rp)$.

Policymakers, military practitioners, and analysts armed with both the dominance equation and the resource expenditure in an adversarial context equation are better armed to understand the internal workings at play during a siege. In turn this can help the onlooked make decisions regarding whether to conduct a siege, how long a siege might last, what to do if under siege, and many other considerations.

Besieged Combatants Options

The besieged combatant has three basic options when faced with how to address the situation they are in. A combatant can (1) solidify the position and conduct a defense, (2) break out of the encirclement and link up with friendly forces, or (3) a friendly force can break in to the encircled perimeter and link up with the besieged combatant. Each of these options is explored in greater detail in the following paragraphs.

Option 1: Solidify the Position and Conduct a Defense (sub-category)

This option is often the response of a combatant who finds themselves under siege in an urban area. The goal of this option is to withstand the besieging force's attacks, while simultaneously increasing the attacker's costs – personnel, equipment, munitions, will, time, and public perception – to the precipice of bankruptcy or the point at which the attacker is no longer will to continue committing resources to the siege. It must be noted that Option 1 can be used in combination with either Option 2 or 3 or be conducted as a unitary operation.

The Islamic State's operations in Mosul from October 2016 to July 2017 are an unsuccessful example of this option. The Islamic State's fighters contracted into the claustrophobic confines of Mosul to both protect themselves from US, Iraqi, and Coalition detection and firepower, but also to retain possession of the city. The Islamic State was successful in driving up the US, Iraqi, and Coalition's costs, but they were not successful in doing so to the point at which the US-Iraqi-Coalition union fell apart or lost sufficient resources to collapse their offensive efforts. Yet, the Islamic State did come close. Many reports emerged during the height of the battle for Mosul that the US was running short on precision guided munitions, thus hampering the US, Iraqi, and Coalition's ability to hasten the Islamic State's end.[62]

Option 2: Break Out and Link Up (sub-category)

Breaking out of an encirclement and linking up with a friendly military force or non-state partner is another option for a besieged actor. This option can be a desperation move because the besieged combatant is running low of resources, a directive from a higher headquarters, or an option selected because the besieged combatant feels a temporal or other situational variable in their favor that makes this move potentially opportune.

Nonetheless, this option entails the besieged combatant penetrating the encirclement, fighting to regain the ability to move on the battlefield, regaining freedom of action, and joining forces with a friendly actor. This option is equally viable in urban terrain, or force-oriented sieges conducted in open terrain. Moreover, this option can also include an air component in which the linkup consists of meeting aircraft to extract the newly liberated force.

As one might imagine, this option requires considerable force and is fairly high risk. Theoretically, a larger penetration effort will require higher resource consumption, but if successful, provide a better avenue for liberated forces to move towards a link up location and friendly force. Yet, a smaller penetration theoretically requires less resources, but if successful, provides a smaller avenue for liberated forces to link up with friend forces. As a result, a larger penetration might take less time but will cost more resources, while a smaller penetration will take less time but cost less resources. The answer to which option is better is purely situational.

The US Army's X Corps operation during the Korean War's Chosin Reservoir campaign (26 November – 13 December 1950) is instructive. The US Army's 7th Infantry Division and US Marine Corps' 1st Marine Division quickly pushed to the Yalu River near the Chinese-North Korean border following General Douglas MacArthur's successful Inchon operation.[63] In X Corps' haste to meet MacArthur's demands to move rapidly, it over-extended itself and as a result the 7th Infantry Division and 1st Marine

---

62 Becca Wasser, et al., *The Air War Against the Islamic State: The Role of Airpower in Defeating ISIS* (Santa Monica, CA: RAND Corporation, 2021), 303-306.
63 Douglas Schaffer, "The Battle of Chosin Reservoir at Yudam-Ni," *Infantry Magazine*, Vol. 92, no. 1 (2003): 17-19.

154 Conflict Realism

Division quick move to the northwest of the Korean peninsula.[64] Chinese forces, who had secretly been supporting North Korean forces, encircled the 7th Infantry Division's 31st Infantry Regiment and the 1st Marine Division at the Chosin Reservoir and subsequently laid siege.[65] The Chinese forces maintained tight control of the perimeter and attempted to annihilate US forces with concentric attacks.[66] Unable to generate a sufficient relief force, X Corps instructed both the 31st Infantry Regiment and the 1st Marine Division to break out of the siege at the Chosin Reservoir and to link up with other X Corps forces further to the east.[67] Vicious fighting ensued. The 31st Infantry Regiment was all but annihilated in its effort to break out of the Chinese perimeter at Chosin Reservoir.[68] On the other hand, the 1st Marine Division fared better, breaking through the Chinese encirclement and linking up with other X Corps elements further east.[69]

Option 3: Break In and Link Up (sub-category)

A break-in and link up by friendly forces is the third option available to a besieged force. This option often comes about in desperate situations, such as when an encircled force faces annihilation if not relieved of encirclement. Furthermore, this option's availability depends on the proximity and freedom of movement of friendly forces, allies, or partners. The goal for the besieged actor in this situation is to hang on long enough for external help. This often entails a defensive operation that seeks to distract the besieging combatant at an area or areas away from the break in point along the encircled perimeter.

Following a successful break in, the besieged combatant has three options. They can defend with the link up force, they can withdraw from the siege, or they can receive material support from the link up force and remain in place. The decision to execute any one of these options is situationally dependent. World War II's Battle of the Bulge is perhaps the most useful example of this option.

During the Battle of the Bulge, the German army encircled US Army forces in and around the Belgium city of Bastogne. Having surrounded

---

[64] Billy Mossman, *Ebb and Flow: November 1950 – July 1951* (Washington, DC: Center of Military History, 1990), 84 – 90.
[65] Mossman, *Ebb and Flow*, 130.
[66] Mossman, *Ebb and Flow*, 136.
[67] Schaffer, "The Battle of Chosin Reservoir at Yudam-Ni," 20.
[68] Mossman, *Ebb and Flow*, 136.
[69] Mossman, *Ebb and Flow*, 147-148.

US forces in Bastogne, the Germans began to concentrically squeeze the Americans.[70] General George Patton and his US 3rd Army quickly responded to the situation at Bastogne. Patton and 3rd Army deftly moved toward Bastogne, penetrated the German's perimeter, and linked up with the besieged US 101st Airborne Division.[71] 3rd Army's effort to link up with the 101st Airborne Division turned the tide of the battle in favor of the US and the Allies.

Conclusion

Sieges are an integral part of war and not some antiquated vestige of warfare from a bygone era. 60 sieges have been identified in war in the post-Cold War period. Nonetheless, a handful of common misconceptions continue to linger over the term siege. Sieges are not a way of warfare, but rather a tool of warfare in a combatant's kit bag. Additionally, sieges can have porous borders and thus the full encirclement of an adversary or object is not necessarily required for an operation to meet the standards to be a siege. Similarly, a siege does not require starvation to be classified as a siege. Starvation can be, and often is, a component of sieges. Moreover, sieges often occur in urban warfare, but sieges are not synonymous with urban warfare. Many types of operations and forms of warfare can occur within urban warfare. Furthermore, the effects of a siege are not limited to the area just under siege. Areas around the siege location are also negatively impacted by the fact that a siege is being conducted in that general area. Basing, support, and firing locations enabling a siege are prone to destruction, collateral damage, and civilian casualties on par with the besieged area.

On the other hand, modern sieges are attritional battles or operations of encirclement and position. Destruction is the currency of a siege and generating submission in an enemy combatant is the siege's ultimate goal. Modern weapons make encirclement relative. If a combatant possesses sufficient firepower in the form of artillery, rockets, missiles, and aerial attack then they can theoretically fulfill an encirclement without having to use land forces to physically surround the targeted combatant.

---

70 Paul Munch, "Patton's Staff and the Battle of the Bulge," *Military Review*, Vol. LXX, no. 5 (1990): 47-48.
71 Munch, "Patton's Staff and the Battle of the Bulge," 50-52.

Sieges are often reactive. Modern sieges tend to not be planned, but rather, they are reactionary actions that a combatant takes to capitalize on a situation. For instance, a military force that fixes another military force in a city, against a large body of water, or restrictive terrain such as woods or mountains, is in an excellent position to conduct a siege. They have pinned their adversary in an area in which it is challenging for the adversary to abscond. Thus, it makes logically sense to tighten the encirclement and begin to either slowly erode the enemy combatant's strength and resources and accelerate the enemy toward exhaustion.

Sieges are not unique to large-scale combat operations or conventional wars. As modern wars have illustrated, sieges cross the boundary from conventional wars to irregular wars and civil wars with ease. The numerous sieges within the Syrian Civil War, for instance, illustrate how sieges can easily move across the various forms of war. Moreover, the siege of Ilovaisk and Donetsk Airport illustrate how irregular forces such as the Donetsk People's Army – a Russian proxy force – can conduct effective sieges in murky proxy wars.

Further, sieges are a game of dominance and resource expenditure in an adversarial context. Dominance is not just a cross check of each combatant's correlation of forces and means. Dominance is about being able to pinch off a piece of a larger operation and intensely focus energy and resources to soundly defeat a combatant in that small window. Trying to dominate everywhere and all the time is unrealistic and a profoundly wasteful endeavor. Understanding the variables and dynamics at work within the idea of dominance is vitally important to correctly assess, apply, and measure success as it relates to trying to dominate an enemy combatant.

Resource expenditure in an adversarial context is closely linked to the idea of dominance. In fact, it is perhaps the most important variable within the range of salient variables. Resource expenditure is not just about having more supplies or people on-hand or close to a significant engagement. Resource expenditure is about identifying the box in which one combatant attempt to dominate another. Within those bounds, identifying the number of contact points and the depth of contact is critical. To be sure, while a front – or contact line – might exist, discrete points of contact exist along those lines. Finding those points and ensuring resource flow to and from those points is how a combatant helps ensure the ability to dominate, or conversely, to not be dominated by an enemy combatant.

The main thing to take away from the analysis of sieges in this chapter is that they will continue to be defining features of war in the future. As

Anthony King reminds us, this is of increasing importance as urban areas continue to grow and military forces – regular and irregular – continue to diminish in size. Sieges and micro-sieges will consume future wars just as they have in wars since the end of the Cold War.

As sieges continue to dominate war, many of the sieges ill effects will continue. This is important for policymakers, military practitioners, and international organizations such as the International Committee of the Red Cross and other humanitarian assistance organizations. Because destruction is the currency of a siege, then the areas in which a besieged force is located will feel the impact of the siege. Therefore, one must expect that forces seeking refuge in urban areas will indirectly invite the death and destruction of civilian populations, civilian infrastructure, and a general disruption to the flow of civilian life. As the Israel-Hamas conflict in 2023-2024 illustrates, sieges in urban areas will also disrupt services such as clean water, medical care, and basic trash removal. One must then expect that disease and many other public health catastrophes will also accompany future conflicts.

In the end, we cannot wish away sieges because we think that they are an antiquated feature of conflict from a bygone area. Sieges are real, tangible, and frightening features of modern – and future – war. Policymakers, military practitioners, analysts, and humanitarian relief organizations must begin to take the siege seriously and develop means and methods through which to account for sieges in future conflicts.

7 On Attrition

Attrition is perhaps one of the most misunderstood and abused ideas in contemporary military thought. Policymakers, military practitioners, and theorists often use and abuse a slew of pejoratives to undercut attrition.[1] This phenomenon is a byproduct of 1980s and 1990s writing which advocated non-attritionalist forms of warfare that appeared to be better aligned to advancing the US Army's AirLand Battle doctrine, Marine Corps warfighting doctrine, and supporting the new All Volunteer Force. The writing and doctrine from this period influenced a generation of military practitioners who are today's senior military leaders and policymakers within the Department of Defense, the United States Government, and many of the US's political-military partners.[2] Many of the assertions made at the time were unscientific, ahistorical, and proffered to generate and maintain consensus for AirLand Battle, yet they continue to resonate deeply with the generation nurtured on those sentiments.

Authors such as William Lind asserted that attrition is a form of warfare.[3] According to Lind, attrition warfare uses firepower, at the expense of movement, to reduce an enemy combatant's numbers. Lind and his coterie of associates further suggest that other types of warfare use firepower and movement to create unexpected and dangerous situations for an adversary.[4] Edward Luttwak takes an almost identical position,

---

[1] Josh Luckenbaugh, "AUSA News: Army Transforming – Not Just Modernizing – for Future Battlefield," *National Defense*, 9 October 2023, accessed 12 December 2023, available at: https://www.nationaldefensemagazine.org/articles/2023/10/9/army-transforming-not-just-modernizing-for-future-battlefield; William Lind, *The New Maneuver Warfare Handbook* (Special Tactics, 2023).

[2] David Johnson, "Shared Problems: The Lessons of AirLand Battle and the 31 Initiatives for Multidomain Battle," *RAND*, April 2018, accessed 12 December 2023, available at: https://www.rand.org/content/dam/rand/pubs/perspectives/PE300/PE301/RAND_PE301.pdf.

[3] William Lind, *Maneuver Warfare Handbook* (London: Routledge, 1985), 4.

[4] See Richard Hook ed., *Maneuver Warfare: An Anthology* (Novato, CA: Presidio Press, 1993).

writing that "an attrition style of war" creates an embellished reliance on firepower at the cost of more movement-centric styles of war.[5]

The commenters of this period thus imply that a dichotomy exists: military forces either use destruction-centric or movement-centric approaches to warfare. Within this dichotomy, movement-centric approaches are high-minded and the zenith of military art, whereas destruction-centric approaches reflect a military force's depravity of mind and practice in the military arts.[6]

The problem with these assertions, however, is that the pragmatic coupling of movement and firepower applies to almost every conceivable type of warfare. One would be hard pressed to find a quality theorist or military (state or otherwise) that does not have the combination of movement, firepower, and surprise at the heart of their approaches to warfare.

Moreover, many of the anti-attrition pejoratives are built on strawmen to advance false information about attrition. As a result, attrition serves as a strawman for policymakers, military practitioners, and theorists to advance self-interested bias and institutional narratives about both war and warfare. What's more, ad hominem is also used to undercut the authority of the individuals who advocate for the usefulness and necessity of destruction-based warfighting in war. Some of the anti-attritionist's comments include referring to those who support destruction-oriented warfare as 'attritionists' or even going so far to suggest these so-called 'attritionists' "don't get it."[7]

Nevertheless, the other side of this discussion finds a handful of contemporary scholars, analysts, and practitioners doing yeoman's work to bridge the gap between the concept's true utility with the animosity and institutional recalcitrance with the concept. These individuals are seeking to reset the discussion and set the record straight on attrition, while chipping away the calcified misinformation surrounding the concept. Analysts Michael Kofman states that attrition, as a matter of historical record, is the

---

5 Edward Luttwak, "The Operational Level of War," *International Security* Vol. 5, no. 3 (1980): 63. doi: https://doi.org/10.2307/2538420.
6 Luttwak, "The Operational Level of War," 63-64.
7 Paul Barnes, "Maneuver Warfare: Reports of My Death Have Been Greatly Exaggerated," *Modern War Institute*, 9 March 2021, accessed 12 December 2023, available at: https://mwi.westpoint.edu/maneuver-warfare-reports-of-my-death-have-been-greatly-exaggerated/; Cole Petersen, "Clearing the Air – Taking Maneuver and Attrition Out of Strategy," *Military Strategy Magazine* Vol. 2, no. 3 (2012), available at: https://www.militarystrategymagazine.com/article/clearing-the-air-taking-manoeuvre-and-attrition-out-of-strategy/.

common way in which wars are waged.[8] In his seminal research project on success in war, *The Allure of Battle: A History of How Wars are Won and Lost*, Historian Cathal Nolan, writes that states are victorious in war as a result of long, bloody, attritional affairs.[9] Scholar Chris Tuck asserts that attrition can be (and is) purposeful because it creates situational and temporal windows of opportunity that pragmatic mobile forces can exploit.[10] Analysts Franz-Stefan Gady and Kofman write is a useful tool when the situation – that is, the disposition, resource availability, time available, among other variables – does not allow a military force to conduct flanking operations, or mobile strikes towards an adversary's rear area.[11] Moreover, scholar Anthony King asserts that destruction-based warfare is all but essential in areas of restrictive terrain, to include urban operating environments.[12] In addition, Mikael Weissman builds upon the ideas of King, correctly pointing out that urban areas are continuing to grow, and therefore, the potential for destruction-oriented fighting in urban areas will increase as we collectively move forward in time.[13]

This chapter examines five of the most prevalent elements of misinformation about attrition. Those five elements include: (1) attrition is a form of warfare, (2) attrition is a correlation of forces and means (COFMs) battle, (3) attrition is focused on a one-to-one exchange ratio between adversaries, (4) attrition abuses one's own logistics, and (5) attrition is a lesser form of warfare. In examining these misunderstandings about attrition, this chapter provides two major findings. First, attrition is not a form of warfare, but a characterization of conflict in which one, or more, makes the pragmatic employment of destruction-based tactics and operations to create or take advantage of tactical and strategic opportunities on the battlefield.

---

8 Michael Kofman, "A Bad Romance: US Operational Concepts Need to Ditch Their Love Affair with Cognitive Paralysis and Make Peace with Attrition," *Modern War Institute*, 31 March 2023, accessed: 13 December 2023, available at: https://mwi.westpoint.edu/a-bad-romance-us-operational-concepts-need-to-ditch-their-love-affair-with-cognitive-paralysis-and-make-peace-with-attrition/.
9 Cathal Nolan, *The Allure of Battle: A History of How Wars Have Been Won and Lost* (Oxford: Oxford University Press, 2017), 577.
10 Christopher Tuck, *Understanding Land Warfare* (London: Routledge, 2022), 99-100.
11 Franz-Stefan Gady and Michael Kofman, "Ukraine's Strategy of Attrition," *Survival* Vol. 65, no. 2 (2023): 7. doi: 10.1080/00396338.2023.2193092.
12 Anthony King, "Urban Insurgency in the Twenty-First Century: Smaller Militaries and Increased Conflict in Cities," *International Affairs* Vol. 98, no. 2 (2022): 621-626. doi: 10.1093/ia/iiac007.
13 Mikael Weissmann, "Urban Warfare: Challenges of Military Operations on Tomorrow's Battlefield," in Mikael Weissman and Niklas Nilsson eds., *Advanced Land Warfare: Tactics and Operations* (Oxford: Oxford University Press, 2023), 148-151.

What's more, it is opportune to recognize that it is time to progress past the use of the word 'attrition' and the use of the phrase 'attrition warfare'. In its place, the defense and security studies community would benefit from identifying exhaustion and force-oriented approaches to warfare as destruction-based approaches. To make this point, this chapter uses this phrase – destruction-based approaches – as a substitute for attrition warfare. Further, it is important to remember that destruction-based warfare is not movement agnostic. Rather, destruction-based approaches are fundamentally grounded in the combination of movement to enable firepower.

Second, one form of warfare does not carry an inherent advantage over another. Rather, forms of warfare organically evolve to the situational requirement(s). As a result, a form of warfare's value resides in its ability to best address the military situation at hand, and not adhere to a state military's preferred doctrine.

Moreover, the forms of warfare, as a rule, correspond to three factors. First, the forms of warfare reflect a state or non-state actor's military goals. If the goal is removing a hostile force from the sovereign territory of another state – like we see with Russia's invasion of Ukraine – then destruction-based warfare is required to push the hostile army out of their neighbor's sovereign territory. On the other side of the coin, if a combatant's goal is a dash to take control of another combatant's capital, then a more movement-centric approach to warfare is required.

Second, the battlefield's situation influences the type of warfare a combatant might employ. A situation can be defined many ways, but in this case a situation includes the physical terrain in which the conflict is occurring, the location of all forces – regular and irregular – throughout the theater of war, the availability of time, and the military objective. A combatant's choice on how they want to fight withers away when weighed against the situation at hand. Thus, the situation has a deterministic effect on campaigns and operations, and subsequently, the tactics therein.

Third, the forms of warfare are reflective of a state's tools of war. A state military heavily invested in a reconnaissance-strike complex and mechanized forces will tend toward a firepower and destruction-based approach to warfighting more so than a state military that cannot support a robust strike and mechanized force. Likewise, non-state actors tend to operate not so much on firepower and destruction, but on movement and making the best use of position.

This chapter proceeds in the following manner. First, the chapter defines attrition. This definition is used to help navigate attrition's elements of misinformation. Through this elements of misinformation, this chapter's thesis – that attrition is not a form of warfare, but rather a characterization of conflict – threads throughout the five assertions. The chapter concludes with a handful of recommendations – reset criteria, if you will – that seek to make understanding were attrition fits within war and warfare clearer.

Defining Attrition

Analyst Trevor Dupuy provides one of the most useful and unbiased examinations of attrition. As a result of his discerning assessment, Dupuy's definition is used as the baseline for what is and is not attrition within this chapter. Dupuy writes that "Attrition is a reduction in the number of personnel, weapons, and equipment in a military unit, organization, or force."[14] Dupuy continues, defining attrition as "The difference between losses and returns to duty." Dupuy does not define attrition as a form of war, but rather, he defines attrition as a result of combat, and therefore as a characterization of warfare in which destruction is the currency and wars focused on exhausting an adversary by increasing the material costs of war beyond what the adversary can sustain.[15] Further, he states that enemy action and accidents are the primary methods through which attrition materializes.

Building on Dupuy's analytical frame, more recent literature describes attrition as a state of being, or put another way, attrition is a characterization, and not a form of warfare.[16] The characterization of 'attrition' can be applied situationally, or generally. For instance, an analyst can describe two tactical forces engaged in destruction-based fighting as a battle of attrition. This term can also apply if one side is using destruction-based methods against their adversary, but not putting their force in situations that allow for a comparable destruction-based approach from their opponent. Further, a combatant might use a destruction-based method, combined with the pragmatic use of terrain, force disposition within the terrain, and timing to avoid having their own force attritted, while inflicting high degrees of destruction on their adversary. This dynamic – the operational and tactical

---

[14] Trevor Dupuy, *Attrition: Forecasting Battle Casualties and Equipment Losses in Modern War* (Falls Church, VA: Hero Books, 1990), 2-3.
[15] Dupuy, *Attrition*, 2-3.
[16] Amos Fox, "Move, Strike, Protect: An Alternative to the Primacy of Decisiveness and the Offense or Defense Dichotomy in Military Thinking," *Association of the United States Army* Land Power Essay 23-4 (2023): 7. Available at: https://www.ausa.org/publications/move-strike-protect-alternative-primacy-decisiveness-and-offense-or-defense-dichotomy.

interplay between a force's location on the battlefield, firepower, and movement – is positional warfare discussed in Chapter 6.[17]

Nonetheless, Tuck notes that some situations require headlong fighting in which both adversarial forces have no recourse, nor method of escape from battering combat.[18] In these instances, in which both forces are engaged in methodical destruction-based warfighting, like the international community witnessed in the latter phase of Operation Inherent Resolve's siege of Mosul, the watchful onlooker can classify this dynamic as a battle of attrition.[19] When combined with the similar dynamic that occurred during the 2015-2016 battle of Ramadi, this campaign can be defined as a war of attrition.[20]

In a conflict in which the entire theater is engulfed in destruction-based warfighting, the war itself can be defined as a war of attrition. Wars of attrition, as Nolan and other scholars remind us, is the womb in which military victory develops.

Examining Attrition's Detractions

Assertion 1: Attrition is a Form of Warfare

Many individuals engaged in the defense and security studies space community imply that attrition is a form, or method, of warfare. This cannot be further from the truth. In a military thinking sense, a 'form', 'method', or 'type' implies that the subject possesses a cohered body of knowledge and a set of operations and tactics. These ideas – the body of knowledge and operations and tactics – might be institutionally developed and maintained, or organically developed by a theorist working outside the confines of an institution. These ideas might be codified as strategy, concepts, or doctrine, if maintained by an institution, such as a Western military force. On the other hand, these ideas might be codified as theory, if they are maintained by scholars, analysts, or theorists.

Nevertheless, an exhaustive examination of open-source Western military strategy, doctrine, and concepts fails to identify any coherent

---

17 Sidharth Kaushal, "Positional Warfare: A Paradigm for Understanding Twenty-First-Century Conflict," *RUSI Journal* Vol. 163, no. 2 (2018): 34-35. doi: 10.1080/03071847.2018.1470395.
18 Tuck, *Understanding Land Warfare*, 99-100.
19 Amos Fox, "The Mosul Study Group and the Lessons of the Battle of Mosul," *Association of the United States Army*, Land Warfare Paper 130 (2019): 4-5.
20 Eric Robinson, et al., *When the Islamic State Comes to Town: The Economic Impact of Islamic State Governance in Iraq and Syria* (Santa Monica, CA: RAND Corporation, 2017), 129-140.

articulation of 'attrition warfare'. That is, none of these institutions possess any semblance of a strategy of attrition, an attritional operating concept, nor a doctrinal framework for attrition warfare and its associated tactics. The US Army's Field Manual 3-0 *Operations* and the British Army's *Land Operations* doctrine are instructive to this point.

Field Manual 3-0 provides only one mention of attrition, and when it does, the purpose is, ironic enough, to assert that attrition is required to achieve victory in war.[21] The British Army's operations doctrine parallels the US Army's absence of a cohered attrition warfare body of knowledge.[22]

Frontal attacks are the closest thing one might find pertaining to attritional tactics in US Army doctrine. Yet, it is important to take a frontal attack in context to the larger picture. Frontal attacks are often not the sole operation or tactic employed in a specific situation but are a component of a larger operation that seeks to enable, collapse, or destroy an adversary through the combination of firepower and movement. Combatants use front attacks to eliminate an adversary's ability to move, and to hold them in place, making them prone to encirclement or destruction. Regrettably, Western military doctrine tends to describe frontal attacks as 'costly' but fails to elaborate on their usefulness in a wide view.

Given the absence in Western military doctrine, as well as defense and security studies or international relations scholarship regarding 'attrition warfare', one must surmise that the word attrition is describing an environment in which destruction is the currency of conflict, and not a form, style, or type of warfare.

Further, a large amount of the literature on forms of warfare suggests that the goal of 'attrition warfare' is to wear an opponent down and outlast them on the battlefield.[23] The problem here is that it is a goal, not a method of warfare. Semantics aside, the differentiation is important. The goal of outlasting an adversary, while preserving one's own combat power is inherent to any actor operating in a competitive environment.[24] Accepting that attrition is an adjective and not a noun, and thus moving forward with a more detailed framework for warfare might well help kickstart the much-needed reset.

---

21 Field Manual 3-0, *Operations* (Washington, DC: Government Publishing Office, 2022): 1-3.
22 Army Doctrine Publication AC 71940, *Land Operations* (Warminster, England), 4-7.
23 Clint Reach, Vikram Kilambi, and Mark Cozad, *Russian Assessment and Applications of the Correlation of Forces and Means* (Santa Monica, CA: RAND Publishing, 2020,,""116.
24 Donella Meadows, *Thinking in Systems: A Primer* (White River Junction: VT, Chelsea Green Publishing, 2008), 16.

Assertion 2: Attrition is a COFMs Battle

Dupuy research finds that "There is no direct relationship between force ratios and attrition rates."[25] According to Dupuy, many factors influence attrition rates. Weather, physical terrain, a force's location, relative combat effectiveness, and many other factors all compete to influence attrition on the battlefield. He concludes his analysis on the subject by positing that neither personnel strength nor force strength ratios impact attrition rates in any meaningful way. Dupuy is not alone in this finding as other researchers have arrived at similar findings stating, among other things, that "Victory in war appears in the main to be determined by factors other than numerical superiority."[26] As a result, when engaged in the exchange of fire, military forces apply multiple types of movement in conjunction with surprise to conserve their force, while attriting that of an adversary.[27]

Wayne Hughes, an operations research analyst who, like Trevor Dupuy, spent a career conducting quantitative analysis on war, writes that destruction-oriented warfare is vital in war. Hughes asserts that in combat, "Victory went to the side whose fire dominated the battlefield… in war the activation and effective employment of superior firepower is the central cause of victory, whether or not casualties determine the outcome."[28] Hughes' emphasis on firepower is the product of firepower's ability to suppress an adversary or keep the adversary's head down and their attention oriented on self-care. According to Hughes, dominant firepower suppresses the enemy; with the enemy focused on self-care, the aggressor can more freely move on the suppressed actor, and consequently apply an inordinate amount of firepower on the adversary (for example, create asymmetric attrition), and obtain victory.[29] Hughes findings on the causality between firepower, suppression, attrition, and battlefield success are quite useful and can be summarized as the following theory:

Victory (v) = dominant firepower (f) → suppression (s) → free movement (m) → asymmetric attrition (a); or, $v = d \rightarrow f \rightarrow s \rightarrow m \rightarrow a$.

---

25 Dupuy, *Attrition*, 98
26 D. Rowland, L.R. Speight, and M.C. Keys, "Manoeuvre Warfare: Some Conditions Associated with Success at the Operational Level," *Military Operations Research* Vol 2, no. 3 (1996): 5.
27 Rowland, Speight, and Keys, "Manoeuvre Warfare," 13-14.
28 Wayne Hughes, "Two Effects of Firepower: Attrition and Suppression," *Military Operations Research* Vol. 1, no. 3 (1995): 30.
29 Hughes, "Two Effects of Firepower: Attrition and Suppression," 30.

166 Conflict Realism

Nonetheless, no compelling or empirical scholarship has emerged to refute Dupuy or Hughes' research. Further, Dupuy's use of attrition in relation to a rate implies its descriptive (adjective) nature, and not a form, method, or style (noun). Considering this chapter's first assertion in relation to Dupuy's proposition, it is safe to say that attrition is not a COFMs battle, but rather, a descriptive term use to describe destruction-oriented warfare.

Assertion 3: Attrition is Focused on a One-to-One Exchange Ratio Between Combatants

This assertion is incorrect on multiple grounds. If attrition is a characterization of war and not actually a method of warfighting, then this assertion's premise is null. Nonetheless, to continue dissecting Assertion 3, let's assume for a moment that attrition is a form of warfare. Let's assume two combatants – Combatants A and B – are both industrialized states and engaged in war with one another. Neither state possesses an asymmetric advantage over the other and thus a degree of parity exists between both combatants. Both combatants are rational actors and each operates according to the logical of economic decision-making and systems theory. Viewed collectively, these ideas form the causal mechanism that drives a military force's form of warfare (see Figure 7.1).

Taking this thought experiment a step further, let's assume that Combatant B is Combatant A's adjacent territorial neighbor. Combatant B has invaded Combatant A. In the process, Combatant B has occupied one-sixth of Combatant A's territory. Diplomacy between the two combatants is at a dead end and military options are Combatant A's only recourse to address the problem.

Militarily, Combatant A has a more open command system in which senior leadership empowers its junior leadership to make on-the-spot decisions. This ethos permeates throughout Combatant A's military force. Combatant B, on the other hand, has a closed command system in which decision-making is slow and centralized. Combatant B therefore operates a command system that is slower, less informed, and less responsive to a current tactical or operational situation than that of Combatant A.

Combatant A intends to use a destruction-based approach against Combatant B. Combatant A intends to operate this way because they understand that Combatant B's armed forces are the virus, so to speak, to which they much eradicate to clarify their political situation and to help nudge Combatant B's policymakers to the negotiation table. Considering that both Combatants A and B are rational actors and economical decision-makers, it is also wise to assume that Combatant B will call for a negotiated end to the conflict at a point far sooner than the outright destruction of their army. Therefore, Combatant A is correct to assume that a destruction-based approach is best for addressing Combatant B. Continuing along this line of causality, it is also correct to assume that Combatant A will likely try to destroy more of Combatant B's materiel at speeds greater than the latter can compensate in order to win the war quicker and more unequivocally.

Combatant A's caveats – avoid large-scale troop deployments and the wanton loss of one's forces and equipment – means that they are not interested in using bad operations, poor tactics, or tic-for-tac casualty exchanges with Combatant B. Combatant A's true military interest is in destroying as much of Combatant B's military force as possible, in the shortest amount of time feasible, while protecting their own force and preventing its destruction.

Despite the clear logic for rational and economic behavior in war, as outline in the preceding thought experiment, history nonetheless provides a instances in which state militaries were forced into situations of relative reciprocation with their adversary. World War II's Eastern Front, for instance, provides many examples in which exchange rates between the Soviet Union's armed forces and those of Nazi Germany were relatively equal.[30] This was more the result of situational factors than preferential methods.

Nevertheless, one would have to eliminate one or more of warfighting's causal mechanisms (Figure 7.1) to assume that Combatant A or B would willingly engage in combat that allowed for a 'one-for-one' exchange rate. At the same time, one would have to assume that a combatant is irrational if it were to remove one or more of the elements of causality. Causality aside, it is dishonest to assume that a state military – fictitious or otherwise – would intentionally operate in an irrational manner, and this is Assertion 3's most egregious leap of logic. States and their militaries do not operate illogically. At least not intentionally.

---

30 David Glantz and Jonathan House, *When Titans Clashed: How the Red Army Stopped Hitler* (Lawrence, KS: University Press of Kansas, 2015), 179-195; Robert Citino, *The Death of the Wehrmacht: The German Campaigns of 1942* (Lawrence, KS: University Press of Kansas, 2007), 96-97.

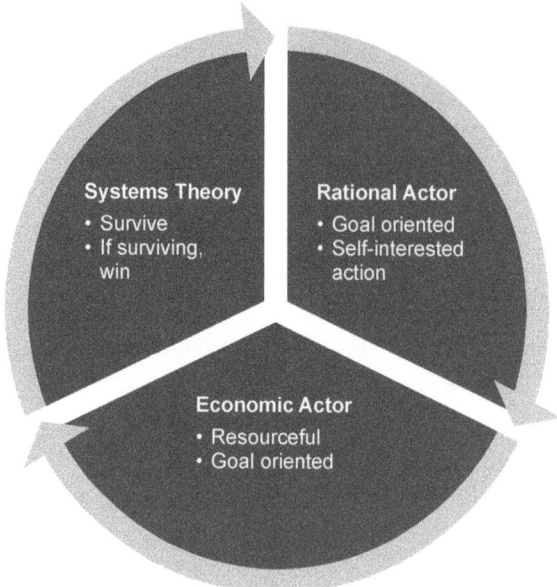

Figure 7.2 Causal Mechanism for a Form of Warfare

Assertion 4: Attrition Abuses One's Own Logistics

Building on the three previous assertions, it is easy to understand that most logistics concerns regarding attrition are unwarranted and over-inflated. The abuse of the logistics argument only stands on merit if one assumes that the combatant using destruction-based warfare is an irrational actor. Yet, we have already established that states and their militaries operate rationally, according to the determinism of systems theory, and economically. To squander one's personnel and equipment through haphazard military operations would be the acme of irrational action. To be sure, the combatant would have to have set aside the prospect of long-term survival – both of the state and their military – to prioritize short-term winning. That is not likely to happen, and states will likely modify their behavior and objectives to achieve balance within their own balancing of systems theory, rationality, and economic thinking.

Assertion 5: Attrition is a Lesser Form of Warfare

Many of the strawmen provided by the late 1970s, 1980s, and 1990s era theorists, continue to erode clear thinking about attrition. Writing in 1979,

Edward Luttwak disparages attrition as a firepower-centric warfare that is out-of-step with the direction the US and NATO should be headed.[31] Luttwak writes:

> We all know what attrition is. It is war in the administrative manner, of Eisenhower rather than Patton, in which the important command decisions are in fact logistic decisions. The enemy is treated as a mere inventory of targets and warfare is a matter of mustering superior resources to destroy his forces by sheer firepower and weight of materiel.

Luttwak offers that more movement-oriented forms of warfare are better than firepower-based forms of warfare. Luttwak provides this opinion without providing empirical evidence to support his argument. Further, he asserts that Western militaries would be best served using an alternative, movement-centric form of warfare, rather than the laborious and synchronized attritional style.[32]

In the mid-1980s, William Lind emerged on the scene as another attrition-detractor. Lind decries attrition as a slow, ponderous approach to warfare that places synchronization, timing, and centralized command and control ahead of responsiveness and surprise.[33] Writing in the early 1990s, John Antal states that armies that adopt an attritional style of warfighting emphasize firepower ahead of movement, and that by doing so, attrition-oriented armies are less capable of inflicting cognitive paralysis on an adversary and winning in a more cost effective manner.[34] Lind, Antal, and Luttwak, theses, in addition to institutional recalcitrance towards the concept's utility, remain today's static which interferes with a clear picture about destruction-oriented warfare.[35]

Many of the points made by individuals such as Luttwak, Lind, and Antal do not stand up to analytical rigor. The empirical work of Hughes, for instance, finds that firepower and destruction are quintessential elements

---

31 Edward Luttwak, "The American Style of Warfare and the Military Balance," *Survival* Vol. 21, no. 2 (1979): 57. doi: 10 1080/00396337908441800.
32 Luttwak, "The American Style of Warfare and the Military Balance," 57-59.
33 William Lind, "The Theory and Practice of Maneuver Warfare," in Richard Hooker ed., *Maneuver Warfare: An Anthology* (Novato, CA: Presidio Press, 1993), 3-6.
34 John Antal, "Thoughts About Maneuver Warfare," in Richard Hooker ed., *Maneuver Warfare: An Anthology* (Novato, CA: Presidio Press, 1993), 67-72.
35 Luckenbaugh, "AUSA News: Army Transitioning – Not Just Modernizing – For Future Battlefield."

of battlefield victory.[36] Moreover, terrain, more so than anything, dictates the speed at which a combatant operates. Terrain further defines whether a military operation or tactical engagement is a headlong clash of forces, or if one combatant is capable of flanking the other combatant and reaching the rear of their formation. Terrain, when combined with an adversary's actions, further complicates matters. An adversary in open terrain might contract into restrictive terrain, such as mountains, dense woods, and urban areas to offset the advantages of a mobile adversary who possesses fire and combat power overmatch.

A combatant's training proficiency is also another factor that determines the swiftness of a combatant's combat operations. To this point, it is also important to convey that combat losses over time change an army. Kofman notes that as a conflict elongates over time, the original, highly trained army of regulars tends to be replaced by hastily trained conscripts.[37] As a result, the combatants both become less adept at synchronized combined arms warfare, and thus, sequential combined arms warfare overtakes the former. It is therefore disingenuous to assert that attrition is a lesser form of warfare. Instead, destruction-oriented warfare often results from necessity.

Further, unless a combatant is fighting a purely defensive war, all combatants are interested in applying the combination of movement and firepower and in generating surprise in an adversary in order to make the most of a tactical engagement or military operation. Even in a defensive war, tactical elements therein are interested in mobility behind their lines so that they can reinforce and conduct other support at various points in their respective lines. As a result, it is disingenuous to suggest that attrition is not an important feature of warfare.

What's more, strategist Alexander Svechin offers that destruction-oriented approaches to warfare are the next logical option when a war cannot be won in a single, decisive strike or battle of annihilation.[38] Svechin writes that destruction-oriented approaches are directed towards obtaining and maintaining material superiority, while depriving a hostile combatant of the means that they need for continued resistance.[39]

---

36 Hughes, "Two Effects of Firepowers," 29-30.
37 Michael Kofman, "Firepower Truly Matters," in *Revolution in Military Affairs* (podcast), 3 December 2023, accessed 12 December 2023, available at: https://shows.acast.com/650105b75a8d440011ecd53c/episodes/firepower-truly-matters.
38 Alexsander Svechin, *Strategy* (Minneapolis, MN: East View Information Services, 1992), 245.
39 Svechin, *Strategy*, 246.

Since history demonstrates that most wars are not won in a singular, decisive strike, it thus makes sense for destruction-oriented operations to take center stage in war.[40] Thus, attrition, although not actually a form of warfare, is not a lesser form of warfighting. Those who make this suggestion are selectively ignoring the impact that deterministic elements such as terrain, time, an adversary's action, and training have on combat.

Conclusion

Attrition is a characterization of conflict – it is an adjective used to provide meaning to engagements, battles, campaigns, operations, and wars in which destruction is high. Moreover, attrition lacks a coherent body of knowledge and an accepted set of practical applications that would allow it to be considered a form of warfare. Therefore, it is prudent to accept that attrition warfare is not actually a typology. Rather it is a misnomer which needs to be rectified. Replacing attrition in all cases in which the defense and security studies community, as well as military practitioners, are not outlining an activity's character is paramount. The term destruction-oriented warfare is an appropriate replacement for attrition's use in regard to a form of warfare.

Further, Western militaries must graduate beyond fanciful and idealist thinking about war. The destruction of hostile armies is how a military creates the situation required for their policymakers to pursue strategic victory. In some instances, however, that is not the case. The threat of, or the bludgeoning push towards the destruction of a hostile army generates the signal for hostile policymakers to negotiate an end to war.

Lastly, if the US remains an economic and military superpower, then it can indeed (and should) operate with a destruction-oriented approach to warfare. When looking for strategic advantage, the US's economic and industrial asymmetry with nearly the rest of the world is one of its most salient, and powerful advantages. It would be foolish to not make the most of that advantage. The US military, and its Western partners, can fight and win large-scale industrial wars in which destruction-oriented combat is the central component. The destruction of armies, or the push toward destroying armies is the most effective, and historically supportable way in which to drive policymakers to the negotiation table.

---

40 Nolan, *The Allure of Battle*, 577.

8   On Precision Strike Strategy

Examining precision strike strategies through the prism of conflict realism results in the identification of a number of logical inconsistencies with contemporary precision strike regimes. To be sure, a fallacy, or paradox, lies at the heart of contemporary precision strike strategies. The belief is that precision strike can reduce civilian harm, protect one's forces, and minimize the amount of explosive munitions required to achieve an effect on a military target. However, the post-9/11 wars clearly illustrate a gap between the theory of precision strike strategy and the practical outcome of precision strike employment. The dissonance between the theory and practical outcome of precision strike strategies can be referred to as the Precision Paradox. That idea is explored in detail within this chapter.

On 4 April 2003, the US Air Force fired multiple Joint Direction Attack Munition (JDAM) satellite-guided bombs at the home of Ali Hassan al-Majid, the cousin of Iraqi President Saddam Hussein, and director of Saddam's intelligence service. Because of his personal relationship with Saddam and his senior position within Saddam's regime, Ali, also known as "Chemical Ali," 'was one of the US coalition's highest-profile targets. In the 'attack's immediate aftermath, US Secretary of Defense, Donald Rumsfeld, stated that "We believe that the reign of terror of Chemical Ali has come to an end."[1]

A few weeks prior, the US made a similar precision-guided munitions attack at 'Dora Farms, a location in which American intelligence thought Saddam Hussein was hiding from the coming US-led invasion of Iraq. Just as at Dora Farms, the attack on Chemical Ali was accurate – precisely hitting the intended target – but they were wholly ineffective. In Saddam's case, the strike was ineffective because he was not present at the time of the attack, whereas with Chemica Ali', he was present, but the attack did not

---

1   "Iraq Missile Attacks Missed Real Targets," *ABC News*, 19 March 2004, accessed 18 August 2023, available at: https://abcnews.go.com/WNT/story?id=131553&page=1.

kill him. Journalists Michael Gordon and Bernard Trainor commented on the attacks,

"Chemical Ali was still alive and well. Saddam and his top officials hand scattered to the four winds, but they were not finished yet."[2]

Summarizing the US's precision strike scorecard between March and April 2003, Douglas Jehl and Eric Schmitt asserted that the US targeted 13 Iraqi senior political and military leaders, but walked away emptyhanded.[3] What's more, the US's precision strike infused decapitation strategy during this period killed over,100 innocent civilians. Unbeknownst at the time, the American score card during this period would serve as a portent of the errancy of the allure of precision strike and so-called "decapitation strategies."[4]

During the US's frenzied departure from Afghanistan in August 2021, American forces – using the latest drone technology – attacked a car that they believed was transporting enemy fighters. The strike was accurate – hitting the car and killing the people inside the car – but it was ineffective because the attack killed noncombatants in the area, including seven children.[5] Azmat Khan, of the New York Times, correctly places this strike in the proper context of both the post-9/11 wars and American precision strike strategy. Khan writes that the civilian's deaths were not the product of a rushed strike conducted under the pressure of a chaotic withdraw from the conflict. Rather, the strike represented one of hundreds, if not thousands, of errant precision strikes, in which a target was hit precisely, but the weapon's destructive impact was felt imprecisely.[6]

Precision strike strategy got a reboot in January 2024. Following troublesome behavior by Yemen's Houthi rebels throughout the Red Sea region, the US and its partners attacked the Houthis with precision strikes.

---

2 Michael Gordon and Bernard Trainor, *Cobra II: The Inside Story of the Invasion and Occupation of Iraq* (New York: Vintage Books, 2007), 523.
3 Douglas Jehl and Eric Schmitt, "The Struggle for Iraq: Intelligence; Errors are Seen in Early Attacks on Iraqi Leaders," *New York Times*, 13 June 2004, accessed 18 August 2023, available at: https://www.nytimes.com/2004/06/13/world/struggle-for-iraq-intelligence-errors-are-seen-early-attacks-iraqi-leaders.html.
4 "Iraq Missile Attacks Missed Real Targets."
5 Azmat Khan, "Military Investigation Reveals How the US Botched a Drone Strike in Kabul," *New York Times*, 6 January 2023, accessed 28 August 2023, available at: https://www.nytimes.com/2023/01/06/us/politics/drone-civilian-deaths-afghanistan.html.
6 Azmat Khan, "The Civilian Casualty Files: Hidden Pentagon Records Reveal Patterns of Failure in Deadly Strikes," *New York Times*, 18 December 2021, accessed 28 August 2023, available at: https://www.nytimes.com/interactive/2021/12/18/us/airstrikes-pentagon-records-civilian-deaths.html

US and partnered precision strikes were a punitive response to the Houthis' relentless attacks on international shipping in the Red Sea region.

However, the troubling trend of accurate, but ineffective strikes continued to pile up. In mid-January 2024, for instance, the *New York Times* reported that 90 percent of the US and its partner's strikes were accurate – that is, they hit their intended target.[7] However, despite that 90 percent accurate rate, nearly three-quarters of the Houthis' military capability remained intact.[8] The situation between the US, its precision strike strategy, and the Houthi rebels in Yemen is another example of the precision paradox at work.

In both Iraq and Afghanistan, a simple pattern emerged around the use of precision strike: (a) strikes are accurate, (b) but the strikes fail to create the targeteer's desired effect, (c) requiring additional strikes to achieve the desired effect, which (d) results in an increase in civilian casualties and collateral damage. The causality that emerges from this exchange of ineffective fire can be appreciated as the '"accurate, but ineffective cycle."' In turn, this cycle facilitates (e) the Precision Paradox theory. Written as a theory, the Precision Paradox's logic can be understood as: *accurate, ineffective precision strike → additional strikes → increase in civilian harm → Precision Paradox*

This theory explains is the causal mechanism that drives the Precision Paradox's recurrence in modern war, and arguably, why the paradox will continue to remain relevant in future wars. Furthermore, the Precision Paradox can be summarized as the incongruence between institutional and industrial-base precision strike narratives and military theories of precision strike strategy, with the objective truth of precision strike's effects.

When examining the Precision Paradox and its relationship to precision strike theory through a conflict realism lens, four foundational points emerge. First, precision strike strategy operates on a flawed logic.[9] That 'flawed logic emerges from the invalidated idea that precision strikes can create sufficient cognitive dislocation to cause an adversarial state, their military, or their military operations to collapse with nary a fight.[10] History has proven this assumption both invalid and fanciful thinking at best.

---

7 Eric Schmitt, "Much of Houthis' Offensive Ability Remains Intact After US-Led Airstrikes," *New York Times*, 13 January 2024, accessed 15 January 2024, available at: https://www.nytimes.com/2024/01/13/us/politics/houthis-yemen-us-airstrikes.html
8 Schmitt, "Much of Houthis' Offensive Ability Remains Intact After US-Led Airstrikes."
9 Precision strike "strategy" also encompasses theory, concepts, and doctrine.
10 Antoine Bousquet, *Scientific Way of Warfare: Order and Chaos on the Battlefields of Modernity* (London: Hurst and Company, 2021), 202-204.

Second, just because a precision strike is accurate, that does not mean that it is effective. As strikes in the 2016-2017 battle of Mosul illustrate, ineffective strikes require additional – not fewer – strikes to accomplish the objective, and or the subsequent use of land force activities to offset the shortcomings of precision strikes.

Third, when precision strikes are ineffective, and additional strikes or land operations are required to create the effect intended with the initial precision strike(s), then precision strategies do not decrease civilian casualties and collateral damage in conflict zones. Instead, battles like the siege of Mosul evince that the cyclic nature of precision strike's accurate, but ineffective attacks, increase civilian casualties and collateral damage.[11] Correspondingly, the Precision Paradox exhibits that as accurate, but ineffective strikes accumulate, so too does civilian casualties and collateral damage.

Fourth, precision strike and precision strike strategy has not hastened a war to a timely conclusion. History does provide, on the other hand, a ledger chock full of wars of attrition precipitated by precision strike strategies. What is certain is that a persuasive argument could be made that applied precision strike strategies do in fact fuel wars of attrition.

Viewed collectively, these four hypotheses are the foundational underpinnings for the Precision Paradox theory. These four principles, and the Precision Paradox collectively, should be the point of departure for anyone interested in examining the voracity of precision strike strategies contemporary or future war.

This chapter proceeds in the following manner. First, the chapter provides background on the evolution of precision strike theory to help clarify how precision warfare has come to represent the supposed zenith of twentieth-century military strategy. This chapter presents a loose pedigree between Napoleon Bonaparte's Austerlitz campaign to today's precision strike warfare to help illustrate precision's importance in military thought. The goal of decisive political-military victory is the glue that binds Austerlitz and precision strike.

Second, this chapter uses three case studies to illustrate the Precision Paradox. The case studies use conflict realist logic to come to their findings. One key element of conflict realism is implications analysis, or the "so what" analysis; that is, just doing X is not enough. Did doing X achieve its

---

11 Becca Wasser, et al., *The Air War Against the Islamic State: The Role of Airpower in Operation Inherent Resolve* (Santa Monica, CA: RAND Corporation, 2021), 168-176.

intended outcome? If so, how and if not, why not? The four case studies in this section use implication analysis to help illustrate precision strike strategy's shortcomings and how the latter fuels the Precision Paradox.

The case studies each illustrate that (1) accurate strikes do not necessarily directly correlate into effective strikes (2) eliminating senior military leadership does not have the impact on operations that precision strike theory supposes, (3) precision warfare often increases civilian casualties and collateral damage, and (4) precision warfare contributes to wars of attrition. This chapter uses the Russo-Ukrainian War to depict that precision strikes on senior military leadership and command nodes does not usher hasten cognitive collapse and accelerate rapid decisive operations, as precision strike theory or precision enthusiasts would have one believe. Next, this chapter probes the Syrian Civil War's battle of Raqqa and Operation Inherent Resolve's siege of Mosul to ascertain that accurate strikes do not correlate to effective strikes. Raqqa and Mosul correspondingly depict precision warfare as a strategy rife with high civilian casualties and collateral damage, while ushering wars of attrition.

Third, as this chapter makes abundantly clear, the Precision Paradox is somewhat deterministic. The deterministic dynamics of the Precision Paradox results in attrition being the paradox's hallmark. Considering the villainy most policymakers and practitioners attribute to attrition, this chapter thus concludes with an assemblage of considerations for anyone interested in precision strike strategy.

Methods and Limitations

The Precision Paradox is a theory which is derivative of an inductive approach to the study of war. The theory is a bridge between precision warfare theory and a general appreciation of modern (and future) war. As noted earlier in this chapter, the theory of Precision Paradox can be summarized simply as: *accurate, ineffective precision strike* → *additional strikes* → *increase in civilian harm* → *Precision Paradox*

The Precision Paradox does not materialize when precision strikes do, in fact, result in first time success, or when a strike achieved its intended purpose with no additional use of force being required on that specific target.

This chapter relies on open-source information to support the veracity of the Precision Paradox. Open-source information is the primary means of information because most quantifiable data on state-based precision strikes is either classified or has not been empirically documented. Nonetheless, this chapter uses qualitative reasoning focused on identifying and mapping causal mechanisms to support the Precision Paradox theory.

This chapter comes with two acknowledged limitations. First is the heavy reliance on US precision theory and usage. That over-reliance is due to the US's overarching lead in drone warfare and PGM use in modern warfare. Second, the theory of the Precision Paradox cannot be supported by quantifiable analysis. For instance, battles such as Raqqa or Mosul cannot be rewound, and the use of precision munitions be replaced with unguided munitions to compare the variances between the two approaches. For this reason, the Precision Paradox is referred to as a heuristic, instead of a law in warfare.

Precision Warfare – Background Information

Antoine Bousquet notes that the allure of precision strike continues to intoxicate political and military leaders with the belief that wars can be won cheaply and quickly, by simply eliminating an adversary's senior leadership with a handful of guided munitions.[12] James Rogers asserts that this idea came to significant prominence with the rise of airpower during the time around World War I.[13]

Nevertheless, precision strike and short-war enthusiasts persist in avoiding evidence to the contrary, and proceed to spin the yarn, positing that selective targeting with precision munitions can induce strategic paralysis and demoralization in an adversary. Paralyzed and demoralized, or so the theory goes, carries with it the potential of triggering the collapse of an adversary's ruling regime, and cause their military to return to their barracks after little to no combat.[14]

The pursuit of precision, put simply, is the quest for accuracy. Accuracy's importance is not an end unto itself, but instead, it is sought-after because of financial prudence.[15] Limited resources constrain all state and non-state actors and, if their resources are not judiciously managed,

---
12 Bousquet, *Scientific Way of Warfare*, 202-204.
13 James Rogers, *Precision: A History of American Warfare* (Manchester, England: Manchester University Press, 2023), 8-9.
14 Christopher Tuck, *Understanding Land Warfare* (London: Routledge, 2022), 50-53.
15 Herbert Van Tuyll and Jurgen Brauer, *Castles, Battles, and Bombs* (Chicago, IL: University of Chicago Press, 2008), 122-132.

hastens their culmination. Andrew Krepinevich, for instance, affirms that efficiency-oriented equipment, organizations, and methods of operating reflect finite means.[16]

Precision theory alleges that accurate weapons systems and operating from stand-off, reduces the number of rounds required to kill or destroy a target.[17] Further, accurate weapons theoretically rouse deft tactical action and consequently hasten battlefield victory.[18] Theorists likewise posit that accurate weapons diminish the strain on an actor's logistics tail.

Throughout the mid- to late-twentieth century, however, humanist perspectives assumed equal, if not greater, importance in warfare than fiscal conservatism. The international community's endorsement of proportionality, military necessity, discrimination, and preventing undue suffering in warfare following World War II reflects this change in attitude.[19] Moreover, the adoption of doctrines, techniques, and training that incorporate international humanitarian law (IHL) are visible reminders of this commitment. The humanist leaning of precision theory appeals to governments interested in maintaining close alignment to IHL.

Modern precision strategies, fall within this context of warfare. Modern precision strategies in the *First Drone Age* use long- and close-range precision weapons systems to support IHL and resource considerations.[20] Additionally, actors use stand-off and smaller land forces to make themselves more challenging to detect and more difficult to engage.

Bousquet posits that many of today's policymakers, scholars, and practitioners make similar pronouncements regarding drone warfare, PGMs, and precision strike strategy's relevance to winning current, and future, war.[21] The 1991 Gulf War, in which the US and its allies routed a vapid Iraqi military, was contemporary precision strike's first real test.[22] During the Gulf War, PGMs accounted for roughly eight percent of expended

---

16 Andrew Krepinevich, *The Military-Technical Revolution: A Preliminary Assessment* (Washington, DC: Center for Strategic and Budgetary Assessments, 2001), 34-38.
17 James Blaker "The American RMA Force: An Alternative to the QDR," *Strategic Review*: 23.
18 David Deptula, *Effects-Based Operations: Change in the Nature of Warfare.* (Arlington, VA: Aerospace Education Foundation, 2001), 11-13.
19 Michael Schmitt, "Precision Attack and International Humanitarian Law," *International Review of the Red Cross* Vol. 87, no. 859 (2005): 462-466. doi: 10.1017/S1816383100184334.
20 Barry Watts, *The Evolution of Precision Strike* (Washington, DC: Center for Budgetary Analysis, 2013), 3-4.
21 Bousquet, *Scientific Way of Warfare*, 117.
22 Malcom Brown, "Intervention That Shaped the Gulf War: The Laser-Guided Bomb," *New York Times*, 26 February 1991, accessed 1 August 2013, available at: https://www.nytimes.com/1991/02/26/science/invention-that-shaped-the-gulf-war-the-laser-guided-bomb.html.

munitions, but their magic accounted for nearly 100 percent of the news coverage in the Western world.[23]

Western military thought and procurement in the post-Gulf War era's,*First Drone Age*' gave rise to an age dominated by claims of precision strike's game changing potential.[24] Western military thought throughout the *First Drone Age* gravitated toward the belief that long-range and standoff PGMs would circumvent land force employment altogether by attacking an adversary's center of gravity, and subsequently, propel the adversary toward psychological paralysis.[25] Military strategists such as US Air Force Colonel John Warden asserted that psychological paralysis would prevent an adversary's ability to develop an effective military response and furnish US military commanders' their own Austerlitz, albeit without the bloody, expensive, and politically dangerous mess of battle and extensive force deployments.[26] Precision strategy, moreover, would encourage a fresh *Revolution in Military Affairs*, which would benefit both the US and its allies.[27] The precision zealotry of this period subsequently motivated several warfighting concepts to include *Rapid Dominance, Parallel Warfare, Effects-Based Operations*, and *Shock and Awe*.

Predictably, PGMs morphed from a niche capability in the early 1990s and into chic new warfighting stratagems by the late 1990s. In 1998's Kosovo War, US General Wesley Clark, NATO's Supreme Allied Commander Europe, recalls that political sensitivities at the time were such that land operations were out of the question, and that NATO would instead fight an air-centric precision strategy in Kosovo.[28] In a campaign almost exclusively fought from the air, US and NATO PGM usage in the Kosovo War reached 29 percent, which was an increase of 21 percent from the Gulf War.[29]

The trend of moving away from land operations toward air-centric precision warfighting quickened during the intervening years. The US invasion of Afghanistan in 2001, for instance, registered 60 percent PGM usage. In 2003,' the US had a 68 percent PGM usage rate.[30] Armed drones and precision technology improved and became cheaper throughout this

---
23 Thomas Mahnken, "Weapons: The Growth and Spread of the Precision Strike Regime," *Daedalus* Vol. 140, no. 3 (2011): 49
24 Mahnken, "Weapons: The Growth and Spread of the Precision Strike Regime," 49.
25 John Warden, "The Enemy as a System," *Airpower Journal* Vol. 9, no. 1 (1995): 41-55.
26 Warden, "The Enemy as a System," 41-55.
27 James Blaker, "The American RMA Force: An Alternative to the QDR," *Strategic Review*, 21-30.
28 Wesley Clark, Waging Modern War (New York: PublicAffairs, 2001), 116.
29 Mahnken, "Weapons: The Growth and Spread of the Precision Strike Regime," 51.
30 Mahnken, "Weapons: The Growth and Spread of the Precision Strike Regime," 51.

period too. This further increased the diffusion of precision warfighting technology and the development of drone-heavy precision-oriented strategy.[31]

Additionally, contemporary conflicts demonstrate that drones, PGMs, and precision strategies are not the antidote for warfare's moral and physical carnage. Scant evidence supports the claims of the precision advocate, such as Harlan Ullman and James Wade, who claim that precision strategies win wars quickly, that they avert battle between land forces, and that precision strategies are less destructive than other warfighting approaches.[32] Facts also fail to sustain the assertion that precision strategies require fewer salvos, fewer munitions, and lessen the logistics load more so than unguided options.[33] The tension between aspirational precision warfare theory and empirical evidence nevertheless permeates the post-9/11. The effect of the tension between precision's aspirational theory and realistic applications in warfare further hastens the Precision Paradox.

The Precision Paradox reveals that the collateral damage, physical destruction, and battlefield death between the *First Drone Age's* PGM-based strategies and those that rely on unguided munitions are commensurable. The Precision Paradox further clarifies that PGM accuracy does not inherently create effectiveness. Namely, a strike might hit its intended target; however, the strike is just as likely to fail to yield its intended effect on the target, despite first hit accuracy. This dynamic – *accurate, but ineffective* strikes – precipitates more strikes to bring about the intended outcome, not fewer. The *accurate, but ineffective* cycle – the Precision Paradox's *élan vital* – makes warfare more destructive and longer lasting, which contradicts precision theory's proclamations of innocuous warfare and hastened wars. Lastly, the Precision Paradox's *accurate, but ineffective dynamic* creates the situation in which precision strategies quickly grind through PGM stocks, surpass industrial base production, and cause campaigns to slow to allow production to catch up, or the combatant uses unguided munitions to bridge the production gap.

---

31 Peter Bergen, Melissa Salyk-Virk, and David Sterman, "World of Drones," *New America*, 30 July 2020, accessed 1 August 2023, available at: https://www.newamerica.org/international-security/reports/world-drones/introduction-how-we-became-a-world-of-drones.
32 Harlan Ullman and James Wade, *Shock and Awe: Achieving Rapid Dominance* (Fort McNair: VA, National Defense University, 1996), 8.
33 Wasser, et al., *The Air War Against the Islamic State*, 305-306.

The problem of supply has shown itself not only in Ukraine, but also in the US-led counter-ISIS campaigns in Iraq, Syria, and The Philippines.[34] The situation is also present as the US, UK, and other partners' fight for control of the shipping lanes in the Red Sea. In this area, the Houthis – an Iranian backed rebel group based in Yemen – continue to attack civilian cargo ships, interrupting international trade. The US, UK, and other partners initiated a precision strike strategy in January 2024 to combat the Houthi rebels. A January 2024 reports posits that the US and partner strikes hit 90 percent of their targets, but that the Houthis retain approximately three-quarters of their ability to fire missiles and employ drones at vessels operating in the Red Sea.[35] Nevertheless, the Precision Paradox, which sees precision theory turned on its head by reality's impact on the praxis of warfare, results in drone warfare and precision-based wars taking the character of slow attritional grinds.

Case Studies

The Precision Paradox is the manifestation of suboptimal and unintended outcomes associated with precision warfare strategies.[36] The Precision Paradox's sub-optimalization arises from the incongruence between aspirational, unrealistic precision warfare theories and the praxis of war. This situation arises because precision thinking heavily relies on linear and optimum scenarios in its development process, while failing to account for the genuine circumstances of war and the acceptance that adversaries are motivated, resourced and learn to adapt while under duress. The Precision Paradox provides several unique features for investigation; however, this chapter limits its focus to (1) civilian casualties and collateral damage, and (2) the financial costs of precision strategies.

A common refrain amongst precision advocates is that precision strategies and PGM use reduces civilian casualties and limits damage to

---

34 Wasser, et al., *The Air War Against the Islamic State*, 305-306; Doug Cameron, "US Struggles to Replenish Munition Stockpiles as Ukraine War Drags On," *Wall Street Journal*, 29 April 2023, accessed 31 August 2023, available at: https://www.wsj.com/articles/u-s-push-to-restock-howitzer-shells-rockets-sent-to-ukraine-bogs-down-f604511a.
35 Eric Schmitt, "Much of Houthis' Offensive Ability Remains Intact After US-Led Airstrikes," *New York Times*, 13 January 2024, accessed 14 January 2024, available at: https://www.nytimes.com/2024/01/13/us/politics/houthis-yemen-us-airstrikes.html.
36 Amos Fox, "The Mosul Study Group and the Lessons of the Battle of Mosul," Association of the United States Army, *Land Warfare Paper 120* (2020): 1-13.

civilian infrastructure.[37] The trope tends to be more factual in counter-terrorism situations, than in more common conflict spaces. Abigail Watson and Alasdair McKay challenge the orthodoxy of precision theory, asserting that the use of PGMs, delivered from long-ranges and from the aerial platforms, prevent an attacking force's ability to implement useful and effective civilian protection mechanisms.[38] Civilian deaths and damage to infrastructure, as a result, do not decrease, but paradoxically increase with the use of drones, long-range fires, and other remote attack systems.

Precision enthusiasts allege that the use of PGM is more economical than the use of unguided munitions, and that PGM use hastens the end of war.[39] War in the *First Drone Age* does not support this assertion. The US's rich use of precision strike technology during Operation Inherent Resolve's battles of Raqqa and Mosul, for instance, each took several months to conclude. Raqqa went on for five months, while Mosul blistered northern Iraq for nearly 10 months. To put Mosul's duration into historical context, World War I's battle of Verdun, one of that war's signature battles, also lasted nearly 10 months, and was dominated by its era's precision strike capability – observed, indirect fire.

The following case studies illustrate the Precision Paradox and those challenges associated with PGMs. This chapter's case studies highlight the causal mechanisms that reinforce the applicability of the Precision Paradox. Resultantly, the case studies provide cautionary tales because they demonstrate what can happen when precision goes wild and does not act in accordance with its theoretical underpinnings.

The Battle of Raqqa (June to October 2017)

In Raqqa, ISIS operated among the city's civilian populace. To protect themselves from adversaries, ISIS used the city's inhabitants as human shields. Similarly, ISIS used the city's infrastructure for protection, basing,

---

[37] "Global Vigilance, Global Reach, Global Power for America" *United States Air Force.* Available at: https://www.af.mil/Portals/1/images/airpower/GV_GR_GP_300DPI.pdf. (Accessed: 30 July 2022).
[38] Abigail Watson and Alasdair McKay, "Remote Warfare: An Introduction," *E-International Relations*, 11 February 2021, accessed 31 August 2023, available at: https://www.e-ir.info/2021/02/11/remote-warfare-a-critical-introduction/.
[39] William Perry, "Perry on Precision Strike," *Air Force Magazine*, 1 April 1997, accessed 30 August 2023, available at: https://www.airforcemag.com/article/0497perry.

and command and control.⁴⁰ ISIS's strategy in Raqqa proved successful – dulling US, coalition, and US proxy efforts to eliminate the organization's stranglehold on the city and its population.

In May 2017, US Secretary of Defense, Jim Mattis, directed a change in US tactics in Syria. US forces operating opposite of ISIS would no longer use attrition. Instead, the US would annihilate ISIS and 'dismantle the ISIS caliphate.'⁴¹ This new focus would begin in the city, thereby denying ISIS's fighters the opportunity to abscond Raqqa.⁴² In practical terms, Mattis' directive meant that US forces would hunt and eliminate ISIS members in Syria, but more importantly, those fighters operating in Syria. Mattis stated that there would be no change to the rules of engagement and that the US would still readily adhere to the international humanitarian law, but his remarks did imply that the US would more ruthlessly pursue ISIS in Raqqa.⁴³

The US's change in strategy and tactics did not result in an increase of US forces on the ground in Syria. Instead, the US relied on the SDF as their proxy for combat operations against ISIS.⁴⁴ The US supported its proxy force with long-range precision fires, armed drones, attack aviation, combat aircraft, and combat advisors, among other support, throughout the battle.⁴⁵ By outsourcing land combat to proxy forces, while maintaining enough advisors to help with targeting and intelligence, the US offloaded the majority of its IHL considerations and responsibility to the SDF.

The US and its coalition and proxy force began the battle of Raqqa in earnest in June 2017. ISIS sought to offset US advantages in sensing, targeting, and precision strike, by operating among the city's civilian populace. ISIS used the city's inhabitants as human shields to flummox, and undermine, US strike capability and to undermine the US's ability to

---

40 Michael McNerney, et al. *Understanding Civilian Harm in Raqqa and Its Implications for Future Conflicts* (Santa Monica, CA: RAND Corporation, 2022) 47-50.
41 Jim Garamone, "Defeat-ISIS 'Annihilation' Campaign Accelerating, Mattis Says," *US Department of Defense*, 28 May 2017, accessed 31 August 2023, available at: https://www.defense.gov/News/News-Stories/Article/Article/1196114/defeat-isis-annihilation-campaign-accelerating-mattis-says/.
42 Garamone, "Defeat-ISIS 'Annihilation' Campaign Accelerating, Mattis Says."
43 Garamone, "Defeat-ISIS 'Annihilation' Campaign Accelerating."
44 Louisa Loveluck and Thomas Gibbons-Neff, "The Islamic State is 'Fighting to the Death' As Civilians Flee Raqqa," *Washington Post*, 8 August 2017, accessed 31 August 2023, available at: https://www.washingtonpost.com/world/middle_east/in-raqqa-a-battle-of-attrition-as-civilians-flee-in-shock/2017/08/07/1e814f9e-78f7-11e7-803f-a6c989606ac7_story.html.
45 "Recommendations to Anti-ISIS Coalition on Operations in Syria," *Center for Civilians in Conflict* (June 2017), 1-2; available at: https://civiliansinconflict.org/wpcontent/uploads/2022/02/CIVICRecsonSyriaJune2017.pdf

protect the civilian population. ISIS also used civilian infrastructure for protection, basing, sustainment, and command and control.[46] ISIS's use of civilian infrastructure unsurprisingly escalated collateral damage and civilian casualties within the city.

ISIS strategy makes sense. ISIS's strategy provided the organization with a method to overcome significant US advantages in remote surveillance, precision strike, and the loss of human capital by confounding the US's ability to discern civilians from fighters.[47] ISIS's strategy and tactics should not come as a surprise. Scholar Michael Schmitt asserts, "When precision capabilities are possessed unequally on the battlefield, the resulting asymmetry may lead the disadvantaged side to resort to tactics that violate the most basic principles of international humanitarian law."[48] ISIS confounded US and coalition attempts to adhere to IHL by operating in ways reminiscent of those experienced by US forces in Iraq during the early phases of Operation Iraqi Freedom.

Further, the US and coalition's use of proxies in close contact with ISIS did not improve the target identification process, which had the second order effect of increasing, not reducing, civilian casualties and damage to infrastructure. US think tank, RAND, seconded this assessment, positing that, "By choosing to conduct the Raqqa operation with a very limited ground presence and a high reliance on the SDF, the United States effectively shifted risk from US military personnel to the civilian population of Raqqa."[49] As a result, the use of PGMs did little to minimize civilian harm during the battle.

The destruction of Raqqa was complete. After the battle, UN advisor Jan Egeland said of Raqqa, "I cannot think of a worse place on earth now."[50] The irony of that statement can be felt when balancing it against the claim of US officials that the conflicts in Syria and Iraq were the most precise bombing campaigns in the history of warfare.[51] The city's population

---

46 McNerney, et al. *Understanding Civilian Harm in Raqqa and Its Implications for Future Conflicts*, 47-50.
47 McNerney, et al. *Understanding Civilian Harm in Raqqa and Its Implications for Future Conflicts*, 91.
48 Schmitt, "Precision Attack and International Humanitarian Law," 466.
49 McNerney, et al. *Understanding Civilian Harm in Raqqa and Its Implications for Future Conflicts*, 91.
50 "No 'worse place on earth' Than Syria's Raqqa, Says Senior UN Adviser Urging Pause in Fighting," *UN News*, 24 August 2017, accessed 31 August 2023, available at: https://news.un.org/en/story/2017/08/563802-no-worse-place-earth-syrias-raqqa-says-senior-un-adviser-urging-pause-fighting.
51 Khan, "Hidden Pentagon Records Reveal Patterns of Failure in Deadly Airstrikes."

was approximately 200,000 prior to the conflict. Roughly 18,000 civilians remained in the city by the battle's end. The fighting within the city destroyed 14 of Raqqa's 24 neighborhoods.[52] Azmat Khan notes that while the number is uncertain, US precision strikes, which often misidentified innocent civilians and civilian infrastructure as hostile targets, significantly contributed to the high degree of civilian deaths during the battle.[53]

The Battle of Mosul (October 2016 – July 2017)

Operation Inherent Resolve's battle of Mosul provides an instructive example for how drone-laden precision strategies in large-scale combat operations precipitate the Precision Paradox. Between October 2016 and July 2017, the combined effect of combat between the US, its coalition, and the Iraqis on one side, and ISIS on the other, milled the city of Mosul. The battle reduced the city's population from two million inhabitants to 700,000 people and created 900,000 internally displaced people (IDP). The battle,destroyed over 40,000 homes, generated more than 10 million tons of rubble, and left Iraq footing a $2 billion reconstruction bill.[54] The nine-month battle was more reminiscent of the hardships of urban battle during World War II than of a high-tech twenty-first Century battle.

Moreover, the US Department of Defense suggests that civilian deaths during the battle range between 320 and 1,400. Several non-government organizations (NGOs), on the other hand, state that the number of civilian deaths ranges between 10,000 – 11,000.[55] Many of these NGOs state their civilian death calculations are the direct result of coalition and Iraqi strikes, and not the result of the combined effect of all the combatants' activity.[56] The *New York Times*, for instance, expounds on the battle's high death toll in

---

52 Loveluck and Gibbons-Neff, "The Islamic State is 'Fighting to the Death' as Civilians Flee Raqqa."
53 Azmat Khan, "The Civilian Casualty Files: Hidden Pentagon Records Reveal Patterns of Failure in Deadly Airstrikes," *New York Times*, 18 December 2021, accessed 31 August 2023, available at: https://www.nytimes.com/interactive/2021/12/18/us/airstrikes-pentagon-records-civilian-deaths.html.
54 Raya Jalabi and Michael Georgy, "This Man is Trying to Rebuild Mosul. He Needs Help – Lots of It," *Reuters*, 21 March 2018, accessed 31 August 2023, available at: https://www.reuters.com/investigates/special-report/iraq-mosul-official.
55 "Iraq: New Report Places Mosul Civilian Death Toll at More Than Ten Times Official Estimates," *Amnesty International*, accessed 31 August 2023, available at: https://www.amnesty.org/en/latest/news/2017/12/iraq-new-reports-place-mosul-civilian-death-toll-at-more-than-ten-times-official-estimates/.
56 "Iraq: New Report Places Mosul Civilian Death Toll at More Than Ten Times Official Estimates."

a 2021 exposé. The report suggests that poor intelligence, hasty targeting, and the demand for results, doomed the US's drone-dominated precision strike strategy in Mosul.

Applying the logic of conflict realism and using implication analysis, it is important to ask the question, why did the US's precision strategy fail to deliver on the concept's theoretical promises? First and foremost was ISIS's basic desire to survive. The desire to survive should come as no surprise when considering basic systems theory and applying a realist's eye to conflict studies. To be sure, theorist Ardant du Picq contends that, "Man in battle…is a being in whom the instinct of self-preservation dominates at certain moments, all other sentiments."[57] Alexander Svechin, echoes du Picq, positing that the first principle in warfare is to insulate oneself against a quick and decisive enemy attack.[58]

ISIS used a variety of tactics, like those they used in Raqqa, in pursuit of self-preservation. ISIS's priority tactic was the avoidance of US, coalition, and Iraqi reconnaissance systems and to confound targeting processes.[59] Tunneling from structure to structure in Mosul was one way that ISIS fighters sought to sidestep reconnaissance and targeting. When PGM strikes were *accurate, but unsuccessful*, ISIS fighters stayed mobile and absconded from the damaged structure to other buildings in the vicinity.[60] ISIS's repositioning drove a new targeting cycle and additional precision strikes.[61]

Second, the iterative character of combat between ISIS and the counter-ISIS forces took the form of a classic *challenge-response cycle*. *Challenge-response cycles* are realist reflections of combat in an adversarial context, which break from monochrome linear theories, like those that dominate precision theory, and demonstrate the true struggle of warfare. *Challenge-response cycles* occur at the micro-level (for example, that of discrete human activity), at the macro-level (for example, that of strategic actor interaction), and all levels of social organization in between. In any

---

57 Ardant du Picq, "Battle Studies: Ancient and Modern Battle" in Greely, J.N. and R.C. Cotton (trans.) *Roots of Strategy: Book 2* (Harrisburg, PA: Stackpole Books, 1987), 77.
58 Alexander Svechin, *Strategy* (Minneapolis, MN: East View Information Services, 1991), 248.
59 Henry Flood, "From Caliphate to Caves: The Islamic State's Asymmetric War in Northern Iraq," *CTC Sentinel* Vol. 11, no. 8 (2018): 30-34.
60 Dan Lamonthe, et al., "Battle of Mosul: How Iraqi Forces Defeated the Islamic State," *Washington Post*, 10 July 2017, accessed 31 August 2023, available at: https://www.washingtonpost.com/graphics/2017/world/battle-for-mosul/.
61 Wasser, et al., *The Air War Against the Islamic State: The Role of Airpower in Operation Inherent Resolve* (Santa Monica, CA: RAND Corporation, 2021), 248-252.

instance, the *challenge-response cycle* continues until one actor either achieves its objective, or it cannot continue competing, and thus, removes itself from the conflict.

Moving from theory to practice, the *challenge-response cycle* was visible in Mosul through the US, the coalition, and the Iraqi's' methodical chase of ISIS with incessant PGM strikes. Despite the US and coalition's best intensions, ISIS's eagerness to survive and win in the face of overwhelming odds resulted in US and coalition PGMs, in conjunction with Iraqi land force operations, tearing Mosul apart as they chased ISIS through the city.[62]

ISIS use of human shields and forcibly relocating civilians to hot spots on the battlefield is another example of the *challenge-response cycle* undercutting precision strike's effectiveness and contributing to the Precision Paradox. The principle of proximity and density dictates that a high proportion of civilians on a battlefield will create a correspondingly high proportion of civilian casualties and collateral damage, regardless of the tactics or munitions used.[63] Viewed collectively, ISIS's use of human shields and forced relocations – the response to the challenge provided by the counter-ISIS forces – accelerated civilian casualties and collateral damage in Mosul.

Illogical tactics, such as 'morale strikes' and 'roof knocking', compounded the inefficiencies already found in precision strike strategy. The Iraqi security forces, for example, were often unwilling to advance on ISIS's fighting positions, whether offensive or defensive. US and coalition forces used morale strikes – attacks on unimportant objects to demonstrate air support's presence – to inspire the Iraqis and compel them to continue advancing.[64] Morale strikes' contribution to civilian casualties and collateral damage in Mosul is unknown, but it certainly contributed to the wave of inefficiency and destruction that swept through the city.[65]

Roof knocking, another example of illogical tactics, is the practice of firing a PGM over a target and detonating the munition overhead. The airburst is intended to create a 'knock'-type effect. The supposed knock is supposed to inspire a building's inhabitants to vacate the facility.[66] Shortly after the knock, a salvo of PGMs follow to eliminate the target. Roof

---

62 Lamonthe, et al., "Battle of Mosul: How Iraqi Forces Defeated the Islamic State."
63 Cathal Nolan, *The Allure of Battle: A History of How Wars Have Been Won and Lost* (Oxford: Oxford University Press, 2017), 370.
64 Wasser, et al., *The Air War Against the Islamic State*, 92-109.
65 Fox, "The Mosul Study Group and the Lessons of the Battle of Mosul," 8.
66 Wasser, et al., *The Air War Against the Islamic State*, 231-232.

knocking in Mosul proved ineffective, wasteful, and destructive because it failed to work, used more munitions when less were required, and created undue damage to civilian infrastructure.[67]

To conclude this case study, open-source information lacks the fidelity to parse the true distinction between which actor bears responsibility for the significant number of civilian casualties and collateral damage in Mosul.[68] Nevertheless, the US's much ballyhooed precision strategy and precision strike capabilities failed to live up to their hype in Mosul. The *challenge-response cycle*, ISIS's unwillingness to be easily and quickly defeated, the Iraqi security forces reluctance to advance in combat, and questionable US and coalition tactics, all contributed to derail the US and coalition's linear precision strategy in Mosul. That strategy fueled paradoxical effects that allowed precision strike to significantly contribute to the creeping wave of attrition in Mosul that disemboweled the city one human life and one man-made structure at a time.

In a brief moment of humility, the US Army acknowledged the limitations of its precision strategy in Mosul. The US Army's Mosul Study Group, a small organization assembled by the US Army's Training and Doctrine Command to distill lessons learn from the battle, asserts that:

> In Mosul, the destruction of physical terrain did not necessarily equate to comparable effects against personnel or communication nodes. Munition choices in Mosul, amplified by the structural density of the city, were not always proportional to the intended effects on the enemy and, when combined with rules of engagement considerations, on collateral damage. Even when considering overpressure and blast waves from these rounds, ISIS fighters were forced from their defensive positions by shrapnel or direct fire weapon systems, rather than blast effects.[69]

---

67 Adam Taylor, "Israel's Controversial 'Roof Knocking' Tactic Appears in Iraq. And This Time, It's the US Doing It," *Washington Post*, 27 April 2016, accessed 31 August 2023, available at: https://www.washingtonpost.com/news/worldviews/wp/2016/04/27/israels-controversial-roof-knocking-tactic-appears-in-iraq-and-this-time-its-the-u-s-doing-it.
68 Wasser, et al., *The Air War Against the Islamic State*, 169.
69 Mosul Study Group, *What the Battle for Mosul Teaches the Force* (Fort Eustis, VA: US Training and Doctrine Command, 2017), 57.

Russo-Ukrainian War

The Russo-Ukrainian War provides a state actor-on-state actor example of the Precision Paradox, whereas Raqqa and Mosul do not. Russia's use of PGMs shows scant difference for how it uses unguided munitions. Russian military tactics in Ukraine appear to be interested in leveling cities and terrorizing civilians and are the primary factor contributing to this situation.[70] Strikes on a maternity hospital in Mariupol (9 March 2022), an attack on a busy train station in Kramatorsk (8 April 2022), a strike on a packed shopping mall in Kremenchuk (27 June 2022), and strikes on a hotel and apartment building in Odesa (1 July 2022) are a few examples of how Russia uses precision to terrorize civilians.[71]

The United Nations' Human Rights Watch reports that as of June 2024, Russia's pitiless siege tactics and indifference for IHL have contributed to more than 30,000 civilian casualties, '10,582 deaths and 19, 875 wounded civilians.[72] Russia's disregard for IHL has also generated seven million IDPs and nearly six million refugees.[73] Russia's brutality, fueled by PGMs, resulted in many world leaders and non-government organizations to classify Russia's actions in Ukraine as war crimes and issue a warrant for the arrest of Russian President Vladimir Putin for violations of international law and fostering a regime dripping with the blood of war crimes.[74]

---

70 John Ismay, "Russian Guided Weapons Miss the Miss, US Defense Officials Say," *New York Times*, 9 May 2022, accessed 31 August 2023, available at: https://www.nytimes.com/2022/05/09/us/politics/russia-air-force-ukraine.html.
71 Daniel Victor and Ivan Nechepurenko, "Russia Repeatedly Strikes Ukraine's Civilians. There's Always an Excuse," *New York Times*, 15 July 2022, accessed 31 August 2023, available at: https://www.nytimes.com/2022/07/02/world/europe/russian-civilian-attacks-ukraine.html.
72 "Two Year Update, Protection of Civilians: Impact of Hostilities on Civilians Since 24 February 2022," United Nations Human Rights Officer of the High Commissioner, https://www.ohchr.org/sites/default/files/2024-02/two-year-update-protection-civilians-impact-hostilities-civilians-24.pdf.
73 "Ukraine: Apparent War Crimes in Russia-Controlled Areas," *Human Rights Watch*, 3 March 2022, accessed 31 August 2023, available at: https://www.hrw.org/news/2022/04/03/ukraine-apparent-war-crimes-russia-controlled-areas.
74 "Ukraine: Apparent War Crimes in Russia-Controlled Areas," *Human Rights Watch*, 3 March 2022, accessed 31 August 2023, available at: https://www.hrw.org/news/2022/04/03/ukraine-apparent-war-crimes-russia-controlled-areas; "Situation of Human Rights in Ukraine in the Context of the Attack by the Russian Federation: 24 February 2022 – 15 May 2022," *Office of the Commissioner United Nations Human Rights*, 29 June 2022, accessed 31 August 2023, available at: https://www.ohchr.org/sites/default/files/documents/countries/ua/2022-06-29/2022-06-UkraineArmedAttack-EN.pdf; Nandita Bose, 'Biden Urges Putin War Crimes Trial After Bucha Killings,' *Reuters*, 4 April 2022, accessed 31 August 2023, available at: https://www.reuters.com/world/biden-says-putin-is-war-criminal-calls-war-crimes-trial-2022-04-04/.

PGM consumption is another consideration to emerge from this conflict and aligns with similar findings from the US's counter-ISIS campaigns in Syria, Iraq, and The Philippines. The bottom line is that PGM utilization is wreaking havoc on all the participants, both active and passive players, and their respective war stocks and industrial bases.[75] Ukraine's supply, greatly enhanced by contributions from Western nations, is churning through precision munitions at a prodigious rate.[76] That problem became more pronounced as Russian forces adapted from early Ukrainian success, which was the result of drone-launched PGMs tank-plinking their way to victory in Kyiv and Kharkiv.[77]

On the other side of the conflict, brutish operations and terror tactics erode Russia's PGM supply.[78] Economic sanctions chastened Russia's ability to import the required components to produce PGMs, challenging Moscow's ability to replace battlefield expenditures.[79] The theft of washing machines and other domestic products by Russian soldiers in Ukraine during the spring of 2022 is symptomatic of the sanctions' impact on PGM production.[80] Pilfered products were shipped back to Russia to help Russian industry offset material shortcomings stifling PGM production.[81] Russia's situation was dire by the summer of 2022, which resulted in Putin turning to Iran for assistance to overcome their PGM deficiency.[82] Douglas Barrie summarizes Precision Paradox-associated challenges presented by

---

75 Jeffrey Sonnenfeld, et al., "Business Retreats and Sanctions Are Crippling the Russian Economy," *SSRN*, 20 July 2022, accessed 31 August 2023, 51-53, available at: https://papers.ssrn.com/sol3/papers.cfm?abstract_id=4167193.
76 Eric Schmitt and Julian Barnes, "Ukraine's Demands for More Weapons Clashes with US Concerns," *New York Times*, 12 July 2022, accessed 31 July 2023, available at: https://www.nytimes.com/2022/07/12/us/politics/ukraine-us-weapons.html.
77 Jack Watling "Ukraine War Update: Dr. Jack Watling," on Arthur Snell (hosted) *Doomsday Watch*, 20 July 2022 [podcast], accessed, available at https://podcasts.apple.com/gb/podcast/war-update-dr-jack-watling/id1593634121?i=1000570572578.
78 Anne Applebaum, "Russia's War Against Ukraine Has Turn into Terrorism," *Atlantic*, 13 July 2022, accessed 31 August 2023, https://www.theatlantic.com/ideas/archive/2022/07/russia-war-crimes-terrorism-definition/670500/.
79 James Byrne, et al., "Silicon Lifeline: Western Electronics at the Heart of Russia's War Machine," *RUSI*, 8 August 2022, accessed 31 August 2023, available at: https://rusi.org/explore-our-research/publications/special-resources/silicon-lifeline-western-electronics-heart-russias-war-machine, p. 5.
80 Jack Detsch, "Ukraine Has Ground Down Russia's Arms Business," *Foreign Policy*, 21 July 2022, accessed 31 July 2023, available at: https://foreignpolicy.com/2022/07/21/ukraine-russia-arms-business-weapons-exports-africa/.
81 Byrne, et al., "Silicon Life," 15-18.
82 Nasser Karimi and Vladimir Isachenkov, "Putin, in Tehran, Gets Strong Supporting Iran Over Ukraine," *Associated Press*, 19 July 2022, accessed 31 August 2023, available at: https://apnews.com/article/russia-ukraine-putin-syria-iran-289c3422c8980e7650dbde2c326d248a.

precision strategies by stating, "The intensity and duration of the Ukraine conflict, and Russia's apparent stockpile issues, will be posing questions for other defense ministries as they reassess their own assumptions on precision weapons stocks and the industrial capacity to replenish them."[83]

Although the conflict is far from over, the Russo-Ukrainian War provides a state-on-state example of the Precision Paradox and illustrates how the paradox resonates with state-on-non-state examples. Scholar Alex Vershinin, for instance, posits that the Russo-Ukrainian War wrecked precision strikes' *one-shot, one-kill* hypothesis, especially as it pertains to large-scale combat operations. Modern militaries are not fragile entities susceptible to defeat in a single decisive strike. Modern militaries reflect a state's adaptive, robust, and redundant system, which pursues both self-preservation and self-interest, in war. Considering modern states' redundancy and resiliency, strategic victory in war today requires inordinate, and unknown, amounts of war materiel and human capital.[84]

Analysis

The analysis here is not focused on completion analysis, but instead this chapter uses implication analysis to provide cogent feedback to help others think more clearly and realistically about the utility and shortcomings of precision strike. Inefficiency is an inherent flaw within precision warfare theory and consequently, inefficiency is one of the driving forces behind the Precision Paradox. Precision warfare's theory is built on a *one shot-one kill* methodology. Yet, theorist Carl von Clausewitz cautions against such unrealistic and linear thinking by stating that, "Man and his affairs, however, are always something short of perfect and will never quite achieve the absolute best."[85]

Clausewitz correctly notes that inefficiency always destabilizes idealistic pursuits of perfection, to which precision warfare theory subscribes. Most precision theory and strategies conflate accuracy with efficiency, and thus bake a critical pitfall into their thinking. However, the reality of the

---

83 Douglas Barrie and Joseph Dempsey, "Russia's Missile Inventories: KITCHEN Use Points to Dwindling Stocks," *IISS*, 11 July 2022, accessed 31 July 2023, available at: https://www.iiss.org/online-analysis//military-balance/2022/07/russias-missile-inventories-kitchen-use-points-to-dwindling-stocks.
84 Alex Vershinin, "The Return of Industrial Warfare," *RUSI*, 17 June 2022, available 31 August 2023, available at: https://rusi.org/explore-our-research/publications/commentary/return-industrial-warfare.
85 Carl von Clausewitz, On War (Princeton: Princeton University Press, 1984), 78.

situation is that high percentage hit rates (for example, accuracy) does not guarantee a high percentage success rate (for example, efficiency).

Inefficient strikes are counterproductive. Recalling the *challenge-response cycle* seen in Mosul, an accurate strike that does not destroy the target, or a strike that fails to achieve its desired effect, can cause the aggressor to re-engage until they achieve their preferred outcome. As the cycle continues, the area and the people touched by each strike's destructive power grows. At the macro-level, the Precision Paradox peaks when the simultaneous cycling of accurate, but ineffective strikes occur at multiple points of contact and along multiple fronts, accelerating the number of civilian casualties, collateral damage, and PGM consumption.

Next, precision advocates assert that precision strategies hasten the duration of war, which in turn, reduces the precisionist's financial stress and strain of warfighting.[86] However, survival instincts and self-interested action unravel precision theory's *one strike, one kill* methodology, resulting in higher, not lower, munition consumption and associated logistics costs.[87]

For instance, Gulf War analysts found that the use of one ton of PGMs in Iraq replaced 12–20 tons of unguided munitions and saved the US 40 tons of fuel per ton of PGM delivered.[88] The Gulf War, however, only represents one data point in the precision warfare era. The post-Gulf War era sees quantitatively higher PGM usage, and this period demonstrates that high first strike hit percentages does not equate to effective strikes, nor do high first strike hit percentages correlate to fiscally responsible warfighting strategies.[89] A team at RAND came to similar conclusions and offered four recommendations to remedy these Precision Paradox-related challenges. First, the West must revamp targeting processes to make it more efficient. Second, the West must modify how it allocates high-demand (for example, PGM) assets. Third, the West must develop useful ways to incorporate unguided munitions to offset PGM depletion. Finally, the West must develop doctrines better suited for PGM employment in urban terrain.[90]

---

86 Perry, "Perry on Precision Strike."
87 Blaker, "The American RMA Force," 23-24.
88 Watts, *The Evolution of Precision Strike*, 8.
89 Michael Pietrucha, "The Five Ring Circus: How Airpower Enthusiasts Forgot About Interdiction," *War on the Rocks*, 29 September 2015, accessed 31 August 2023, available at: https://warontherocks.com/2015/09/the-five-ring-circus-how-airpower-enthusiasts-forgot-about-interdiction/.
90 Wasser, et al., *The Air War Against the Islamic State*, 306-307.

Conclusion

Conflict realism finds that the Precision Paradox is both a cautionary tale and a self-correction mechanism. Implication analysis helps illustrate that the Precision Paradox provides notice that military thought must move evolve from fanciful beliefs in unproven and historically inaccurate postulates about technology and warfare. Modern drone enthusiasts, such as James Rogers, are taking notice. Rogers cautions that as the US loses agency as both a drone and precision hegemon, drone-driven atrocities will likely increase.[91]

The Precision Paradox rises from precision theories and strategies lacking a requisite degree of realism. Antoine Jomini notes that this is not a new challenge in the practice of warfare, stating, "Correct theories, founded upon right principles, sustained by actual events of war, and added to accurate military history, will form a true school of instruction for generals."[92] J.F.C. Fuller echoes Jomini by stating that, "Method creates doctrine, and a common doctrine is the cement which holds an army together…we want the best cement, and we shall never get it unless we can analyze war scientifically and discover its values."[93] Clausewitz offers perhaps the most apropos, and well-known, argument for realism in military thought, positing that warfare is a human endeavor, and as a result, fog, friction, and chance will always impact even the most meticulous plans.[94]

To help overcome the Precision Paradox in future warfare, practical, empirical information must be the cornerstone for military thought – from conceptual work to doctrines, to policy and strategy, and tactical plans.

Military thought must move beyond the pursuit of the modern-day equivalent of 'decisive battle' through unrealistic precision strategies. Modern actors, embodied by dynamic systems, are too redundant and robust for such an approach to work. Instead, modern, and future military thought must possess realist verities of war and warfare, account for the *challenge-response* cycle, and move beyond the unrealistic belief that *one-shot,*

---

91 James Rogers, "Drone Warfare: The Death of Precision," *Bulletin of the Atomic Scientists*, 12 May 2017, accessed 31 August 2023, available at: https://thebulletin.org/2017/05/drone-warfare-the-death-of-precision/.
92 Antoine Jomini, "The Art of War," in J.D. Hittle (ed.) *Roots of Strategy: Book 2* (Harrisburg, PA: Stackpole Books, 1997), 556.
93 J.F.C. Fuller, The Foundations of the Science of War (London: Hutchinson and Company, 1926), 35.
94 Clausewitz, *On War*, 101.

one-kill theories are truly viable. The failure to do so will perpetuate the Precision Paradox and its devastating battlefield effects.

Furthermore, Western militaries must move beyond the advocacy of narrative at the expense of reality." The reliance on narrative at the expense of reality – in which precision strike is often no less deadly for noncombatants or civilians as are ballistic munitions – will continue to needless spread the Precision Paradox.

Conclusion

The analytical layer between international relations scholarship and military doctrine is underrepresented in contemporary conflict studies. International relations scholarship tends to focus on the interaction amongst states and how and why they go to war. Most Western military thought, on the other hand, focuses far too tactically and on advancing narratives at the cost of providing objective analysis. Conflict realism provides an alternative method and angle through which to analyze armed conflict. Conflict realism looks to find causality and use that explain the phenomenon of war and warfare.

In this book, we have used conflict realism as a lens to examine war and warfare with the goal of coming to a handful of findings that define modern war, and perhaps might offer portends for the future of armed conflict. Conflict realism is the layer that separates international relations scholarship and international relations theory from military scholarship, and military doctrine. Conflict realism is a form of military theory that looks to the future not based on hope, aspiration, or fictitious imagination. Conflict realism identifies causality and causal mechanisms, as well as those with enduring qualities, to make forecasts for the future. Moreover, conflict realism carries forward long standing trends in war and warfare and assumes that persistent features of armed conflict will continue to persist well into the future. For instance, the deterministic impact of terrain on military operations is something that technology will not be able to circumvent, obviate, or otherwise neutralize. Terrain will always impact how forces operate on the ground and how air forces and how airborne sensors and air launched munitions can affect land forces – both regular and irregular – operating in complicated terrain. Conflict realism works to bring the reality of conflict to the fore, while still making useful and relevant predictions about the future of armed conflict.

It is easy to get caught up in the hype about future war. To be sure, many so-called futurists are making hay over the seeming dominance that drones and artificial intelligence will play in future armed conflict.¹ Yet, most of their analysis is based on unrealistic assessments of war that are missing context. For instance, analysis based largely on YouTube and TikTok videos of drone strikes from Nagorno-Karabakh and Ukraine provide an unrealistic depiction of drone capability.² They do so by only showing drone success and they do not show the number of drone failures, nor do they anticipate the comparable developments in anti-drone technology that will inevitably come along in the near future. Moreover, drone enthusiasts oftentimes see drones as a unique form of warfare, while failing to appreciate that they are little more than just another wrinkle of combined arms and joint warfare.³ Further, as Mike Kofman notes, cyber capabilities are just now reaching the point in which they are effective at interdicting precision strike munitions and drones in Ukraine.⁴ Thus, one can expect the knowledge to escape that theater of conflict and begin to populate in militaries and non-state forces throughout the world in the coming years. In doing so, the perceived dominance of precision strike and drone technology will dissipate as more and more military forces and non-state actors invest in cyber capabilities to combat the guidance, operating, and command and control systems directing precision strike and drones.

In addition, it is easy to get caught up in the belief that social media's impact on war and warfare will be significant. Social media's impact on armed conflict largely remains in the cognitive space, but as the wars in Ukraine and Gaza have illustrated, targeted uses of social media can be used to rouse international support for a national cause. Social media in the early days of Russia's 2022 reinvasion of Ukraine, for instance, helped galvanize Western support for Ukraine's cause. Social media posts showing Ukrainian civilians resisting against Russian armed forces dominated the early days of the conflict and helped inspire Ukrainian support across the world.⁵ Likewise, social media posts of Hamas' atrocities in their October

---

1 John Antal, "Next War Webinar John Antal," *Fight Club International*, 29 January 2024, accessed 21 March 2024, available at: https://www.youtube.com/watch?v=yoktqS7aWgo.
2 John Antal, "Azerbaijan and Armenia," *Maneuver Warfighter Conference*, 7 March 2022, accessed 21 March 2022, available at: https://www.youtube.com/watch?v=_At9txsUKIw&rco=1.
3 Amanpour, "Experts on Ukraine Drone Attacks: 'Technology Will Never Take Away the Grim Reality of War'," *CNN*, 31 August 2023, accessed 21 March 2024, available at: https://www.cnn.com/videos/tv/2023/08/31/amanpour-shirreff-rogers-ukraine-drones.cnn.
4 Michael Kofman, "Russia and Ukraine with Mike Kofman," *Revolution in Military Affairs* (podcast), 11 April 2024.
5 Megan Specia, "'Like a Weapon': Ukrainians Use Social Media to Stir Resistance," *New York Times*, 25 March 2022, accessed 21 March 2024, available at: https://www.nytimes.com/2022/03/25/world/europe/ukraine-war-social-media.html.

2023 invasion of Israel illustrated how social media works in the other direction.⁶ The videos of Hamas foot soldiers invading and terrorizing Israel galvanized support for Israel's punitive incursion into Gaza, which in 2024 remains a hotly debated subject. Further, in recent years, books have been published discussing how war will be transformed by social media and its ability to mobilize support for a cause, provide tactical battlefield directions, and issue strategic guidance, amongst many other things.⁷

Innocently operating through intermediaries in armed conflict is another element of war and warfare that contemporary commenters like to posit will characterize future war. Language such as 'by, with, and through,' is used to disguise what is really proxy war strategy. Moreover, proxies are often dressed in more appealing and respectful language, being referred to as partners, and in some cases allies. Thus, many commenters suggest that Western militaries can distance themselves from the rigors of armed conflict by operating by, with, and through partners – often local actors – in pursuit of self-interest in a specific conflict. Yet, most of these arguments fail to account for the agency costs that are associated by working with a proxy pursuant to one's objectives, and therefore, they fail to capture the range of agency costs that a principal-proxy relationship entails. Agency costs, which are unique to a relationship, generally play a deterministic role in shaping the strategic relationship between a principal and its proxy in these situations.

The reliance on proxy forces, moreover, fuels long and destructive wars of attrition. This is because a state's use of proxy strategies disaggregates the Clausewitzian trinity, which asserts that the passion of a state's people is one of the conditions which help govern – for example, control – how a policymaker conducts war. Yet, by detaching the warfighters from the state and outsourcing that requirement elsewhere, the passion of the people no longer possesses any sort of regulatory balance on policymakers charged with waging war. Thus, wars can be left to destructively burn on for years and years at a time, while societies at home feel little impact of these policy decisions. The United States' war in Afghanistan, a 20-year boondoggle that ended in humiliation and defeat, is an excellent example of this situation in modern armed conflict.

---

6 Desiree Abid, "Amid Israel-Hamas Conflict, 'Information War' Plays Out on Social Media, Experts Say," *ABC News*, 24 November 2023, accessed 21 March 2024, available at: https://abcnews.go.com/International/social-media-information-war-israel-hamas-conflict/story?id=104845039.
7 See P.W. Singer and Emerson Brooking, *Like War: The Weaponization of Social Media* (Boston: Houghton Mifflin Harcourt, 2018); John Antal, *Next War: Reimaging How We Fight* (Philadelphia: Casemate Publishers, 2023); David Patrikarakos, *War in 140 Characters: How Social Media is Reshaping Conflict in the Twenty-First Century* (New York: Basic Books, 2017).

In addition, sieges, of which 60 unique examples have been identified since the end of the Cold War, are a defining feature of armed conflict in this strategic environment. Sieges frequently occur in modern wars because of the asymmetry between military forces, with the weaker force seeking refuge in an urban area to offset its materiel inferiority. Further, the vast amounts of overhead sensors – to include space-based capabilities and drones – in addition to cyber intelligence, linked with long-range precision fires has rendered operating in the open nearly suicidal. Since all actors are intent first on survival and second on obtaining their military objectives, operating in and from the city is one of the best ways in which to offset many of the sensors, cyber, and precision effects on the modern battlefield. As a result, sieges inevitably develop.

Where does this leave us? A few deductions can be drawn. First, the land domain remains the dominant domain in which wars are fought. Other domains, such as the air, sea, space, and cyber domains, are all just facilitators of land warfare. Land warfare still remains supreme in war because war is a continuation of policy by other means, and policy directly corresponds to political institutions and the states in which they govern. Driving an adversarial state toward a policy change requires the application of hard power in a tangible way on, as Clausewitz noted, either their force, their 'capital', or their allies. Operating at these strategic focal points is most effectively done through the application of land power, on the land domain, with support through combined arms and joint capabilities coming through complementary domains.

Many of the individuals hyping drones and long-range precision strike are not seeing those things for what they are, which is little more than observation and attack from above. Viewed as 'observations and attacks from above', and not as a novel individual weapon system allows the analyst to see that many of the ballyhooed pieces of new technology are just additional adjuncts of combined arms and joint warfare. Little more.

Observations and attacks from above generate a basic survival response in any combatant operating in an open, adaptive system, which is why one sees the growth of tunnels, trenches, and bunkers on battlefields in Ukraine, Gaza, Iraq, and Syria. Just as soldiers did in World War I, fighters go underground when overhead sensors are airborne and when artillery, rockets, and missile fire dominate the battlefield.

Viewed collectively, the continued integration of new technology which seeks more efficient ways to kill one's enemy while protecting one's own forces, with another combatant's self-interest in survival and equally winning the conflict results in paradoxical results. Wars most often become attritional affairs exactly because they want to efficiently kill their opponent, while remaining protected from destruction themselves, which creates micro-cycles of violence across battlefields.

Each of these micro-cycles of violence represents its own fight for a contested piece of territory – whether for the territory itself, or to destroy the enemy force that happens to be located on that territory. In the aggregate, one can then say that land wars are fought for territorial possession. As the history of war and warfare has illustrated, there is no-game changing technology in war or warfare, nor is there any game-changing tactics. Conflict realism cautions that as new technology is introduced, the side introducing the technology will have a fleeting window of asymmetric advantage over their adversary. During that period, they can (and should) make the most of the opportunity at hand and attempt to get the great gain toward their strategic and political aims. Simultaneously, the other combatant will soon realize their predicament and respond to overcome their opponent's advantage. Moreover, they will likely seek to develop comparable advantages at this time so that they can introduce challenges into the conflict that provide them fleeting advantage too.

From a conflict realist position, that leaves us at a basic position as it pertains to war and warfare. War on the land domain will remain the most important place in which wars are fought. All other domains and all other types of warfighting support warfighting on the land domain. Finally, conflict realism finds that the challenges of land warfare are enduring features of war and warfare, regardless of technological innovation.

Across the course of this book's research, seven challenges of land warfare emerged. First, armies must be capable of taking or retaking territory. The use of the word 'armies' here is a generic term for any land force – state army, non-state military force, or any other irregular actor. Taking or retaking territory is an enduring challenge of land warfare because it is the central element of warfare – armies invade other states to take territory; invaded states use their armies to fight invading armies; non-state forces attack government forces to gain political advantage. This is the most basic building block in war and warfare.

Second, conflict realism dictates that armies must be capable of clearing armies from a specific territorial holding. As we have examined thus far, combatants generally do not elect to fight one another in the open, prone to the advantages that each other can bring to bear on the other. They seek to operate from terrain and in ways that enhances their own capabilities and technology while undercutting that of their opponent. This mentality means that inspired armies are likely to be found dug into fortified positions and ready to defend themselves or counterattack as required. Considering the fact that aerial bombardment has limited effects on entrenched forces, and to retake territory will require the clearance of contested ground with land forces, then it must be understood that future land forces should be large, armored, and resilient. You do not want a land force to culminate, for example, exhaust its offensive or defensive capability, as it cleared an objective of enemy forces. If that occurs, it is increasingly prone to counterattack, which leads to the next challenge.

Armies must be capable of holding territory. It is not enough to be able to clear a hostile force from a specific piece of territory, but armies must be capable of holding that territory. What that means is that they must not culminate in the process of taking or retaking terrain, or they must not fritter away all their combat power in the process of claiming or reclaiming some piece of land. Doing so makes an army prone to counterattack, and thus quickly losing the territory that they likely fought hard to gain.

The focus conflict realism places on urban warfare and sieges makes it apparent that armies must be capable of encircling a hostile force. Encirclements are important tactics and operations in modern wars because that are directly related to system manipulation, and thus, strategic victory. Encirclement allows a combatant to control the flow of resources to and from an enemy combatant, which in turn allows them to accelerate their enemy toward strategic materiel exhaustion, which is the catalyst for strategic military defeat.

Further, an army must be capable of holding a hostile force in position. This allows the army to more effectively integrate joint effects against an enemy combatant. Moreover, being capable of holding a hostile force also allows one to account for most of the other challenges already outlined.

Finally, armies must be capable of sealing boundaries. As Ukraine's war with Russia illustrates, if a state's force cannot maintain the integrity of their international borders, then that state's sovereignty might always be called into question by a hostile neighbor. Hamas' invasion of Israel in October 2023 makes the same point apparent. However, this challenge is equally a post facto point. Armies must be capable of taking or retaking territory, clearing enemy forces from desired territory, and holding desired territory. Sealing boundaries, to include international boundaries, should be viewed in conjunction with those other challenges, almost as an adjunct that helps complete the foundation upon which the other challenges build.

Conflict realism thus concludes with a handful of recommendations for Western militaries that are interested in thoughtful innovation for future armed conflict. First, larger, not smaller, land forces are required to address the challenges of battlefield transparency, precision strike, and drone usage. This recommendation is at odds with most of the commentary dominating Western military discourse today. Most Western military discourse provides an emotionally charged and reactionary response to the challenges of battlefield transparency, precision strike, and drones. The response is to get smaller, lighter, and disperse. On the surface, this makes sense, but this does not take a deeper examination of how small, lighter, and dispersed will equally fair when paired against transparency, precision strike, and ubiquitous drones. The argument can also be made, which has been made within this book, that small and lighter forces operating in a dispersed manner against those threats will be in a far more dangerous position to identification, monitoring, and destruction than larger and more resilient forces would otherwise be. Therefore, Western militaries should invest in larger, more armored armies because doing so will allow them to fight to a military objective, overcome whatever resistance that they find in their way, and still possess sufficient combat power to stave off culmination. Moreover, this will allow armies to possess sufficient combat power to hold an objective and parry concerted enemy counterattacks aimed at overturning the former's gains.

Conflict realism also demonstrates that joint services – air forces, navies, etc. – remain a supplement element in war and warfare and the wars are still won and lost on land. Therefore, states must remain focused on the strategical alignment within their armed forces. Air forces and navies inherently have more technologically advanced warfighting systems and expensive budgets than armies, but armies possess the implements that have historical shown to win wars time and time again – soldiers. Therefore, conflict realism posits that states must remain strategically aligned with what wins their wars – it is not their air forces, their navies, or any other service, it is their army. Armies are the force that can protect the state's territory. Armies are the force that fulfill foreign policy goals. Armies are the force that protect civilian populations and partner with and assure allies and partners.

Bibliography

Books

Antal, J. The Next War: Reimaging How We Fight. Philadelphia: Casemate Pubishers, 2023.

Antal, J. *Seven Seconds to Die: A Military Analysis of the Second Nagorno-Karabakh War and the Future of Warfighting.* Havertown, PA: Casemate Publishers, 2022.

Beevor, A. *Stalingrad, The Fateful Siege: 1942-1943.* New York: Penguin Books, 1998.

Billingsley, D., and Grau, L. *Fangs of the Lone Wolf: Chechen Tactics in the Russian-Chechen Wars, 1994-2009.* Fort Leavenworth, KS: Foreign Military Studies Offices, 2012.

Black, J. *A Short History of War.* New Haven, CT: Yale University Press, 2021.

Bousquet, A. *Scientific Way of Warfare: Order and Chaos on the Battlefields of Modernity.* London: Hurst and Company, 2021.

Brauer, J. and Van Tuyll, H. *Castles, Battles, and Bombs: How Economics Explains Military History.* Chicago: University of Chicago Press, 2008.

Chandler, D. *Napoleon.* London: Pen and Sword Publishers, 2000.

Chandler, D. *The Art of Warfare in the Age of Marlborough.* New York: Sarpedon, 1990.

Citino, R. *The Death of the Wehrmacht: The German Campaigns of 1942.* Lawrence, KS: University Press of Kansas, 2022.

Citino, R. *The Wehrmacht's Last Stand: The German Campaigns of 1944-1945.* Lawrence, KS: University of Kansas Press, 2017.

Clausewitz, C. *On War.* Princeton: Princeton University Press, 1986.

Clark, W. *Waging Modern War.* New York: PublicAffairs, 2001.

Coll, S. *Directorate S: The CIA and America's Secret War in Afghanistan and Pakistan.* New York: Penguin Press, 2018.

Dupuy, T. *Attrition: Forecasting Battle Casualties and Equipment Losses in Modern War.* Fairfax, VA: HERO Books, 1990.

Dupuy, T. *Understanding War: History and Theory of Combat.* London: Leo Cooper, 1987.

Dodge, B. and Grau, L. *Fangs of the Lone Wolf: Chechen Tactics in the Russian-Chechen Wars, 1994-2009* (Fort Leavenworth, KS: Foreign Military Studies Offices, 2012.

Dolman, E. *Pure Strategy: Power and Principle in the Space and Information Age* (London: Frank Cass, 2005.

Fuller, J. *Generalship, Its Diseases and Their Cure: A Study of the Personal Factor in Command.* Harrisburg, PA: Military Services Publishing Company, 1936.

Fuller, J. *The Foundations of the Science of War.* London: Hutchinson and Company, 1926.

Fuller, J. *Training for Soldiers*. London: Hugh Rees, 1914.
Glantz, D. and House, J. *When Titans Clashed: How the Red Army Stopped Hitler*. Lawrence, KS: University Press of Kansas, 2015.
Glantz, D. and House, J. *To the Gates of Stalingrad: Soviet-German Combat Operations, April-August 1942*. Lawrence, KS: University Press of Kansas, 2009.
Glaser, C. *Rational Theory of International Politics: The Logic of Competition and Cooperation*. Princeton, NJ: Princeton University Press, 2010.
Greely, J.N. and Cotton, R.C. (trans.) *Roots of Strategy: Book 2*. Harrisburg, PA: Stackpole Books, 1987.
Godfroy, J., Powell, J., Morton, M., & Zais, M. *US Army in the Iraq War – Volume 2: Surge and Withdrawal, 2007-2011*. Carlisle Barracks, PA: US Army War College Press, 2019.
Gordon, M, and Bernard T. *Cobra II: The Inside Story of the Invasion and Occupation of Iraq*. New York: Vintage Books, 2006.
Hayes, J. *Combat Stories Map: A Historical Repository and After Action Tool for Capturing Stories, Storing, and Analyzing Georeferenced Individual Combat Narratives*. Monterey, CA: Naval Postgraduate School, 2016.
Hollis, M. *The Philosophy of Social Science: An Introduction*. Cambridge: Cambridge University Press, 1994.
Howard, M. *War in European History*. Oxford: Oxford University Press, 2009.
Hooker, R., ed. *Maneuver Warfare: An Anthology*. Novato, CA: Presidio Press, 1993.
Hymel, K. *Patton's War: An American General's Combat Leadership*, Volume 2, August-December 1944. Columbia, MO: University of Missouri Press, 2023.
Isbell, B.R. "The Future of Surprise on the Transparent Battlefield." In Brian Holden Reid ed., *The Science of War: Back to First Principles*. London: Taylor and Francis Group, 1993.
Kedzior, Richard. *Endurance and Evolution: The US Army Division in the Twentieth Century*. Santa Monica, CA: RAND Corporation, 2000.
King, A. *Urban Warfare in the Twenty-First Century*. Cambridge, England: Polity Publishing, 2021.
Knezys, S., and Sedickas, R. *The War in Chechnya*. College Station, TX: Texas A&M University Press, 1999.
Liddell Hart, B.H. *The Ghost of Napoleon*. London: Faber and Faber Limited, 1934.
Lind. W. *The New Maneuver Warfare Handbook*. Livermore, CA: Special Tactics, 2023.
Lind, W. *Maneuver Warfare Handbook*. London, England: Routledge, 1986.
McNerney, M. et al. *Understanding Civilian Harm in Raqqa and Its Implications for Future Conflicts*. Santa Monica, CA: RAND Corporation, 2022
Martini, J. et al., *Operation Inherent Resolve: U.S. Ground Force Contributions*. Santa Monica, CA: RAND Corporation, 2022.
McFate, S. *The New Rules of War: Victory in the Age of Durable Disorder*. New York: William Marrow, 2019.
Meadows, D. *Thinking in Systems: A Primer* (White River Junction, VT: Chelsea Green Publishing, 2008.
Mearsheimer, J. *The Great Illusion: Liberal Dreams and International Realities*. New Haven, CT: Yale University Press, 2015.

Mearsheimer, J. *The Tragedy of Great Power Politics.* New York: W. W. Norton & Company, 2001.

Mossman, *Ebb and Flow: November 1950 – July 1951.* Washington, DC: Center of Military History, 1990.

Patrikarakos, D. *War in 140 Characters: How Social Media is Reshaping Conflict in the Twenty-First Century.* New York: Basic Books, 2017

Robinson, E. et al., *When the Islamic State Comes to Town: The Economic Impact of Islamic State Governance in Iraq and Syria.* Santa Monica, CA: RAND Corporation, 2017.

Rogers, J. *Precision: A History of American Warfare.* Manchester, England: Manchester University Press, 2023.

Rosenau, J. "Thinking Theory Thoroughly," in Rosenau, J. and Durfee, M. eds *The Scientific Study of Foreign Policy.* London: Routledge, 2018.

Ullman, H. and J. Wade. *Shock and Awe: Achieving Rapid Dominance.* Washington, DC: National Defense University, 1996.

Schelling, T. *Arms and Influence.* New Haven, CT: Yale University Press, 1966.

Simpkin, R. *Race to the Swift: Thoughts on Twenty-First Century Warfare.* London, England: Brassey's, 1985.

Svechin, A. *Strategy.* Minneapolis, MN: East View Information Services, 2004.

Tuck, C. *Understanding Land Warfare.* London, England: Routledge, 2022.

Oliker, O. *Russia's Chechen Wars 1994-2000, Lessons from Urban Combat.* Monterey, CA: RAND Corporation, 2001.

Patton, G. *War as I Knew It.* New York: Houghton Mifflin, 1995.

Rayburn, J., & Sobchak, F. *The US Army in the Iraq War – Volume 1: Invasion – Insurgency – Civil War, 2003-2006.* Carlisle Barracks, PA: US Army War College Press, 2019.

Ullman, H. and Wade, J. *Shock and Awe: Achieving Rapid Dominance.* Fort McNair: VA, National Defense University, 1996.

Singer, P.W. and Brooking, E. *Like War: The Weaponization of Social Media.* Boston: Houghton Mifflin Harcourt, 2018.

Schom, A. *Napoleon Bonaparte: A Life.* New York: HarperCollins Publishers, 1997.

Paret, P. *The Cognitive Challenges of War, Prussia 1806.* Princeton: Princeton University Press, 2009.

Wasser, B., et al. *The Air War Against the Islamic State: The Role of Airpower in Defeating ISIS.* Santa Monica, CA: RAND Corporation, 2021.

Watling, J. *The Arms of the Future: Technology and Close Combat in the Twenty-First Century.* London: Bloomsbury Publishing, 2024.

Watling, J. and Reynolds, N. *Meatgrinder: Russian Tactics in the Second Year of Its Invasion of Ukraine.* London: RUSI, 2023.

Watling, J. and Reynolds, N. *War by Others' Means: Developing Effective Partner Force Capacity Building.* London: RUSI, 2021.

Watts, B. *The Evolution of Precision Strike.* Washington, DC: Center for Budgetary Analysis, 2013.

Weigley, R. *The Age of Battles: The Quest for Decisive Warfare from Breitenfeld to Waterloo.* Bloomington, IN: Indiana University Press, 1991.

Weissman, M. and Nilsson, N. eds., *Advanced Land Warfare: Tactics and Operations*. Oxford: Oxford University Press, 2023.

Journal Articles

Altman, D. "Advancing without Attacking: The Strategic Game Around the Use of Force," *Security Studies* (Vol 27, no. 1, August 2018).

Altman, D. "By Fait Accompli, Not Coercion: How States Wrest Territory from Their Adversaries," *International Studies Quarterly*, Vol. 61 (2017).

Andersen, J. "The Paradox of Precision and the Weapons Review Regime," *Philosophical Journal of Conflict and Violence* Vol. 1 (2017).

Blaker, J. "The American RMA Force: An Alternative to the QDR," *Strategic Review*: 23.

Carr. A. "It's About Time: Strategy and Temporal Phenomena," *Journal of Strategic Security*, Vol. 44, no. 3 (2021).

Deptula, D. *Effects-Based Operations: Change in the Nature of Warfare*. (Arlington, VA: Aerospace Education Foundation, 2001).

Flood, H. "From Caliphate to Caves: The Islamic State's Asymmetric War in Northern Iraq," *CTC Sentinel* Vol. 11, no. 8 (2018).

Fox, A. "Move, Strike, Protect: An Alternative to the Primacy of Decisiveness and the Offense or Defense Dichotomy in Military Thinking," *Association of the United States Army* Land Power Essay 23-4 (2023).

Fox, A. "On Sieges," *RUSI Journal* Vol. 166, no. 2 (2021).

Fox, A. "Maneuver is Dead? Understanding the Components and Conditions of Warfighting." *RUSI Journal* Vol. 166, no. 6-7 (2021).

Fox, A. "The Siege of Ilovaisk: Manufactured Insurgencies and Decision in War," *Association of the United States Army*, Land Power Essay 21-2 (2021).

Fox, A. "Getting Multidomain Operations Right: Two Critical Flaws in the US Army's Multidomain Operations Concept," *Association of the United States Army*, Land Warfare Paper 133 (2020).

Fox, A. "The Mosul Study Group and the Lessons of the Battle of Mosul," *Association of the United States Army*, Land Warfare Paper 130 (2020).

Fox, A. "On the Employment of Cavalry," *ARMOR* 123, no. 1 (Winter 2020).

Fox, A. "'Cyborgs at Little Stalingrad': A Brief History of the Battles of Donetsk Airport, 26 May 2014 to 21 January 2015," *Institute of Land Warfare*, Land Warfare Paper 125 (2019).

Fuller, J. The Foundations of the Science of War; Gold Medal (Military) Prize Essay for 1919. *RUSI Journal* Vol. 65, no. 458 (1920).

Gady, F. and Kofman, M. "Ukraine's Strategy of Attrition," *Survival* Vol. 65, no. 2 (2023).

Hecht, E. "Defeat Mechanisms: The Rationale Behind the Strategy." *Military Strategy Magazine*, Vol. 4, no. 2 (2014).

Hoffman, F. "Defeat Mechanisms in Modern Warfare." *Parameters* 51, no. 4 (2021).

Hughes, W. "Two Effects of Firepower: Attrition and Suppression," *Military Operations Research* Vol. 1, no. 3 (1995).

Kaushal, S. "Positional Warfare: A Paradigm for Understanding Twenty-First-Century Conflict," *RUSI Journal* Vol. 163, no. 2 (2018).

King, A. "Will Inter-State War Take Place in Cities?" *Journal of Strategic Studies* Vol. 45, no. 1 (2022).

King, A., "Urban Insurgency in the Twenty-First Century: Smaller Militaries and Increased Conflict in Cities," *International Affairs*, Vol. 98, no. 2 (2022).

Lider, Julian, "The Correlation of World Forces: the Soviet Concept," *Journal of Peace Research* Vol. 17, no. 2 (1980).

Luttwak, E. "The Operational Level of War," *International Security* Vol. 5, no. 3 (1980).

Luttwak, E. "The American Style of Warfare and the Military Balance," *Survival* Vol. 21, no. 2 (1979).

Mahnken, T. "Weapons: The Growth and Spread of the Precision Strike Regime," *Daedalus* Vol. 140, no. 3 (2011).

Meiser, J. "Are Our Strategic Models Flawed? Ends + Ways + Means = (Bad) Strategy." *Parameters* Vol. 46, no. 4 (2016).

Mumford, A. "Proxy Warfare and the Future of Conflict," *RUSI Journal* Vol. 158, no. 2 (2013).

Munch, P. "Patton's Staff and the Battle of the Bulge," *Military Review*, Vol. 70, no. 5 (1990).

Owen, W. 'The False Lessons of Modern War: Why Ignorance is Not Insight,' *The British Army Review*, Vol. 183 (2023).

Petersen, C. "Clearing the Air – Taking Maneuver and Attrition Out of Strategy," *Military Strategy Magazine* Vol. 2, no. 3 (2012).

Porter, P. "Out of the Shadows: Ukraine and the Shock of Non-Hybrid War," *Journal of Global Security Studies*, 8(3) (2023).

Schaeffer, D. "The Battle of Chosin Reservoir at Yudam-Ni," *Infantry Magazine*, Vol. 92, no. 1 (2003).

Schmitt, M. "Precision Attack and International Humanitarian Law," *International Review of the Red Cross* Vol. 87, no. 859 (2005).

Schmitt, Olivier. "Wartime Paradigms and the Future of Western Military Power." *International Affairs* 96, no. 2 (2020).

Starry, D. "To Change an Army." *Military Review* 43, no. 3 (March 1983).

Warden, J. "The Enemy as a System," *Airpower Journal* Vol. 9, no. 1 (1995).

Government Publications

Field Manual 3-0, *Operations*. Washington, DC: Government Printing Office, 2017.

Joint Publication 3-0, *Joint Campaigns and Operations*. Washington, DC: Government Printing Office, 2023.

Reports and Research Papers

"Russian, Ukrainian Bases Endangering Civilian," *Human Rights Watch*, 21 July 2022.

"Ukraine: Apparent War Crimes in Russia-Controlled Areas," *Human Rights Watch*, 3 March 2022.

"Situation of Human Rights in Ukraine in the Context of the Attack by the Russian Federation: 24 February 2022 – 15 May 2022," *Office of the Commissioner United Nations Human Rights*, 29 June 2022.

"Military Satellites by Country 2023," *World Population Review*, accessed 1 September 2023.

"Convention (IV) respecting the Laws and Customs of War on Land and its annex: Regulations concerning the Laws and Customs of War on Land," *The Hague*, 18 October 1907.

"Protocol Additional to the Geneva Conventions of 12 August 1949, and relating to the Protection of Victims of Non-International Armed Conflicts (Protocol II), *United Nations*, 8 June 1977.

"Rome Statute of the International Criminal Court," *United Nations*, 17 July 1998.

"Rule 53, Starvation as a Method of Warfare, Customary International Humanitarian Law," *International Humanitarian Law Database*.

"General Assembly Overwhelmingly Adopts Resolution Demanding Russian Federation Immediately End Illegal Use of Force in Ukraine, Withdraw All Troops," *United Nations*, 2 March 2022.

"General Assembly Adopts Resolution Calling for Immediate, Sustained Humanitarian Truce Leading to Cessation of Hostilities Between Israel, Hamas," *United Nations Tenth Emergency Special Session*, 40th and 41st Meetings, 27 October 2023.

"Fact Sheet: One Year of Supporting Ukraine," *White House*, 21 February 2023.

"Joint Statement on Israel," *White House*, 9 October 2023.

"Global Vigilance, Global Reach, Global Power for America" *United States Air Force*.

"NATO's Response to Russia's Invasion of Ukraine," *NATO*, 6 November 2023.

Mosul Study Group, *What the Battle for Mosul Teaches the Force*. Fort Eustis, VA: US Training and Doctrine Command, 2017.

"Iraq: New Report Places Mosul Civilian Death Toll at More Than Ten Times Official Estimates," *Amnesty International*, December 2017.

"Recommendations to Anti-ISIS Coalition on Operations in Syria," *Center for Civilians in Conflict*, June 2017.

Antal, J. "The First War Primarily with Unmanned Systems: Ten Lessons from the Second Nagorno-Karabakh War," 2021.

Antal, J. "Seven Battlefield Disrupters: Warfighting Challenges for the US Military Derived from the Second Nagorno-Karabakh War," *Maneuver Warfighter Conference*, February 2022.

Barrie, D. and Dempsey, J. "Russia's Missile Inventories: KITCHEN Use Points to Dwindling Stocks," *IISS*, 11 July 2022.

Evans, A., et al., "Russian Offensive Campaign Assessment," *Institute for the Study of War*, 27 December 2023.

Garamone, J. "Defeat-ISIS 'Annihilation' Campaign Accelerating, Mattis Says," *US Department of Defense*, 28 May 2017.

Ingram, H. and Mohammed, O. "The Logic of Insanity: Why Groups Like ISIS and Hamas Strategically Court with Self Destruction," *George Washington University Program on Extremism*, 1 November 2023.

Krepinevich, A. *The Military-Technical Revolution: A Preliminary Assessment* (Washington, DC: Center for Strategic and Budgetary Assessments, 2001.

Light, P. "Rumsfeld's Revolution at Defense," *Brookings Institute*, Policy Brief #142, 2005.

Montgomery, A. and Nelson, A. *The Rise of the Futurists: The Perils of Predicting with Futurethink* (Washington, DC: Brookings Institution, 2022).

Obama, B. "The Way Forward in Afghanistan," *Obama White House Archives*, accessed 4 November 2023.

Roberts, B. "On Theories of Victory, Red and Blue." Lawrence Livermore National Laboratory, Livermore Papers on Global Security No. 7 (2020).

Roberts, B. "On Theories of Victory, Red and Blue," *Lawrence Livermore National Laboratory*, Livermore Papers on Global Security No. 7 (2020): 26-39.

Stapleton, B. "The Problem with the Light Footprint: Shifting Tactics in Lieu of Strategy," *CATO Institute*, 7 June 2016.

Sonnenfeld, J., et al., "Business Retreats and Sanctions Are Crippling the Russian Economy," *SSRN*, 20 July 2022.

Newspaper Articles and Websites

"AUSA Coffee Series – GEN James Rainey – US Army Futures Command," *Association of the United States Army*, 14 December 2023.

"Interview with Lieutenant Colonel James Rainey," DVIDS, 16 November 2004.

"Ukraine Blew Up a dam to Stop the Russian Advance on Kyiv, Some Homes Remain Flooded," *Reuters*, 28 May 2022.

"Wagner Group Prison Recruits Back in Russia from Ukraine Front Lines Accused of Murder and Sexual Assault," *CBS News*, 27 June 2023.

"Iraq Missile Attacks Missed Real Targets," *ABC News*, 19 March 2004.

"The Azerbaijan-Armenia Conflict Hints at the Future of War," *The Economist*, 8 October 2020.

"US Struggles to Replenish Munition Stockpiles as Ukraine War Drags On," *Wall Street Journal*, 29 April 2023.

Abid, D. "Amid Israel-Hamas Conflict, 'Information War' Plays Out on Social Media, Experts Say," *ABC News*, 24 November 2023.

Adams, P. "Ukraine War: Why Russia's Infrastructure Strikes Strategy Isn't Working." *BBC News*, 9 March 2023.

Amanpour, C. "Experts on Ukraine Drone Attacks: 'Technology Will Never Take Away the Grim Reality of War'," *CNN*, 31 August 2023.

Antal, J. "Learning from Recent Wars – Observations from the Second Nagorno-Karabakh War and the Russian Ukrainian War, *European Security and Defence*, 5 October 2022.

Antal, J. "Next War Webinar John Antal," *Fight Club International*, 29 January 2024.

Antal, J. "Azerbaijan and Armenia," *Maneuver Warfighter Conference*, 7 March 2022.

Applebaum, A. "Russia's War Against Ukraine Has Turn into Terrorism," *Atlantic*, 13 July 2022.
Babchenko, A. "The Savagery of War: A Soldier Looks Back at Chechnya," *Independent*, November 10, 2007.
Barnes, P. "Maneuver Warfare: Reports of My Death Have Been Greatly Exaggerated," *Modern War Institute*, 9 March 2021.
Barno, D. and Benasahel, N. "The Other Big Lessons That the U.S. Army Should Learn from Ukraine," *War on the Rocks*, 27 June 2022.
Bergen, P. Salyk-Virk, M., and Sterman, D. "World of Drones," *New America*, 30 July 2020.
Bose, N. 'Biden Urges Putin War Crimes Trial After Bucha Killings,' *Reuters*, 4 April 2022.
Bousquet, A. "The Battlefield is Dead," *Aeon*, 9 October 2017.
Burns, R. "Russia's Failure to Take Down Kyiv was a Defeat for the Ages," *Associated Press*, 7 April 2022.
Bush, G. "The Surge of Troops in Iraq," *PBS*, 4 May 2020.
Brown, M. "Intervention That Shaped the Gulf War: The Laser-Guided Bomb," *New York Times*, 26 February 1991.
Bremlow, B. "A Brief, Bloody War in a Corner of Asia is a Warning About Why the Tank's Days of Dominance May Be Over," *Business Insider*, 24 November 2020.
Byrne, J., et al. "Silicon Lifeline: Western Electronics at the Heart of Russia's War Machine," *RUSI*, 8 August 2022.
Detsch, J. "Ukraine Has Ground Down Russia's Arms Business," *Foreign Policy*, 21 July 2022.
Dewan, A. "Ukraine and Russia's Militaries are David and Goliath. Here's How They Compare," *CNN*, 25 February 2022.
Doyle, G., Rao, A., and Kawoosa, V. "Shaping the Battlefield: How Weapons from Western Allies are Strengthening Ukraine's Defense," *Reuters*, 10 March 2023.
Dzhulay, D. "Revealed: How Ukraine Blew Up a Dam to Save Kyiv," *RadioFreeEurope / RadioLiberty*, 26 February 2023.
Eckel, M. "The Bakhmut Meat Grinder: Russian Troops Are Pummeling This Donbas City. It's Unclear Why," *RadioFreeEurope/RadioLiberty*, 13 December 2022.
Eversden, A. "Wormuth: Here are the 6 Areas the Army Must Prepare for in 2030." *Breaking Defense*, 15 September 2022.
Finnegan, C., et al. 'Pentagon Expects to Deploy Troops to Mideast After Iraqi Protesters Assault US Embassy, Torch Guardhouse to Protest Airstrikes', *ABC News*, 1 January 2020.
Fox. A. and Kopsch, T. "Moving Beyond Mechanical Metaphors: Debunking the Applicability of Centers of Gravity in 21st Century Warfare, *The Strategy Bridge*, 2 June 2017.
Fox, A. "Urban Warfare, Sieges, and Israel's Looming Invasion of Gaza." *War on the Rocks*, 27 October 2023.
Fox, A. "The War for the Soul of Military Thought: Futurists, Traditionalists, Institutionalists, and Conflict Realists," *Association of the United States Army*, 17 March 2023.

Gatopoulos, A. "The Nagorno-Karabakh Conflict is Ushering in a New Age of Warfare," *Al Jazeera*, 11 October 2020.

Gardner, F. "Ukraine War: Is the Tank Doomed?," *BBC News*, 7 July 2022.

Gladstone, R. "UN Says Islamic State Executed Hundreds During Siege of Mosul," *New York Times*, 2 November 2017.

Gordon, M., Coles, I. and Adnan, G. 'Trump Vows Retaliation After Iranian-Backed Militia Supporters Try to Storm U.S. Embassy in Baghdad', *Wall Street Journal*, 31 December 2019.

Hambling, D. "The 'Magic Bullet' Drones Behind Azerbaijan's Victory Over Armenia," *Forbes*, 10 November 2020.

Hasik, James. "Precision Weapons Revolution Changes Everything." *CEPA*, 17 February 2023.

Hutchinson, B. "How Humanitarian Corridors Work to Offer Lifeline to Besieged Ukrainians," *ABC News*, 12 April 2022.

Ismay, J. "Russian Guided Weapons Miss the Miss, US Defense Officials Say," *New York Times*, 9 May 2022.

Jalabi, R. and Georgy, M. "This Man is Trying to Rebuild Mosul. He Needs Help – Lots of It," *Reuters*, 21 March 2018.

Jehl, D. and Schmitt, E. "The Struggle for Iraq: Intelligence; Errors are Seen in Early Attacks on Iraqi Leaders," *New York Times*, 13 June 2004.

Johnson, D. "Shared Problems: The Lessons of AirLand Battle and the 31 Initiatives for Multidomain Battle," *RAND*, April 2018.

Karimi, N. and Isachenkov, V. "Putin, in Tehran, Gets Strong Supporting Iran Over Ukraine," *Associated Press*, 19 July 2022.

Khan, A. "Military Investigation Reveals How the US Botched a Drone Strike in Kabul," *New York Times*, 6 January 2023.

Khan, A. "The Civilian Casualty Files: Hidden Pentagon Records Reveal Patterns of Failure in Deadly Strikes," *New York Times*, 18 December 2021.

Kingsley, P. and Bergman, R. "The Secrets Hamas Knew About Israel's Military," *New York Times*, 13 October 2023.

Kofman, M. "A Bad Romance: US Operational Concepts Need to Ditch Their Love Affair with Cognitive Paralysis and Make Peace with Attrition," *Modern War Institute*, 31 March 2023.

Lamonthe, D., et al. "Battle of Mosul: How Iraqi Forces Defeated the Islamic State," *Washington Post*, 10 July 2017.

Loveluck, L. and Gibbons-Neff, T. "The Islamic State is 'Fighting to the Death' As Civilians Flee Raqqa," *Washington Post*, 8 August 2017.

Luckenbaugh, J. "AUSA News: Army Transforming – Not Just Modernizing – for Future Battlefield," *National Defense*, 9 October 2023.

Matisek, J. and Bertram, I. "The Death of American Conventional War: It's the Political Willpower, Stupid," *The Strategy Bridge*, 5 November 2017.

McKew, M. "The Gerasimov Doctrine: It's Russia's New Chaos Theory of Political Warfare and It's Probably Being Used on You." *Politico*, 5 September 2017.

al-Mughrabi, N., & Rose, E. Israeli Troops Raid Gaza as Arab Nations Condemn Bombardment. *Reuters*. 26 October 2024.

Nissenbaum, D., Lubold, G., Lieber, D., & Abdel-Baqui, O. Israel Agrees to US Request to Delay Invasion of Gaza. *Wall Street Journal*, 25 October 2023.

Peck, M. "Losses in Ukraine Are 'Out of Proportion' to What NATO Has Been Planning For, the Alliances Top General Says," *Business Insider*, 5 February 2023.

Perry, W. "Perry on Precision Strike," *Air Force Magazine*, 1 April 1997.

Pietrucha, M. "The Five Ring Circus: How Airpower Enthusiasts Forgot About Interdiction," *War on the Rocks*, 29 September 2015.

Rogers, J. "Drone Warfare: The Death of Precision," *Bulletin of the Atomic Scientists*, 12 May 2017.

Shalal, A., & Singh, K. Biden, Key Western Leaders Urge Israel to Protect Civilians. *Reuters*, 22 October 2023.

Schwirtz, M. "Last Stand at Azovstal: Inside the Siege That Shaped the Ukraine War," *New York Times*, 24 July 2022.

Schmitt, E. "Much of Houthis' Offensive Ability Remains Intact After US-Led Airstrikes," *New York Times*, 13 January 2024.

Schmitt, E. and Barnes, J. "Ukraine's Demands for More Weapons Clashes with US Concerns," *New York Times*, 12 July 2022.

Specia, M. "'Like a Weapon': Ukrainians Use Social Media to Stir Resistance," *New York Times*, 25 March 2022.

Spencer, J., Collins, L. and Geroux, J. "Case Study #7 – Fallujah II," *Modern War Institute*, 25 July 2023.

Stewart, P. "Exclusive: US Assesses Up to 60% Failure Rate for Some Russian Missiles, Officials Say." *Reuters*, 25 March 2022.

Taylor, A. "Israel's Controversial 'Roof Knocking' Tactic Appears in Iraq. And This Time, It's the US Doing It," *Washington Post*, 27 April 2016.

Triebert, C. "Video Reviews How Russian Mercenaries Recruit Inmates for Ukraine War," *New York Times*, 16 September 2022.

Vershinin, A. "The Return of Industrial Warfare," *RUSI*, 17 June 2022.

Victor, D. and Nechepurenko, I. "Russia Repeatedly Strikes Ukraine's Civilians. There's Always an Excuse," *New York Times*, 15 July 2022.

Watson, A. and McKay, A. "Remote Warfare: An Introduction," *E-International Relations*, 11 February 2021.

Williams, R.F.M. "The Rise and Fall of the Pentomic Army." *War on the Rocks*, 25 November 2022.

Podcasts

Gady, F. "What if the Deep Battle Doesn't Matter?", *This Means War* (podcast), 14 September 2023.

Kofman, M. "Russia and Ukraine with Mike Kofman," *Revolution in Military Affairs* (podcast), 11 April 2024.

Kofman, M. "Firepower Truly Matters," in *Revolution in Military Affairs* (podcast), 3 December 2023.

Langston, H. "Reflects on Russia's 2014-2015 Donbas Campaign," *Revolution in Military Affairs* (podcast), 4 March 2024.

Spencer, J. "The Battle of Sadr City, March-May 2008," *Urban Warfare Project* (podcast), 6 March 2020.

Watling, J. "Ukraine War Update: Dr. Jack Watling," *Doomsday Watch* (podcast), 20 July 2022.

Index

Afghanistan (US war, 2001-2021) 11, 21, 36, 45, 57, 69, 115, 147, 173, 179, 197
Altman, Daniel 109
Antal, John 169
Attrition 40, 62-63, 80, 86, 118-121, 127, 147, 159-171, 176
    Wars of attrition 13, 27, 40, 42, 99
    Attritional warfare 11, 85, 164-165
Autonomous system 2, 8, 37, 55, 59, 61, 64, 83, 112, 117, 173, 180, 183, 185-186, 193

Barno, David 60
Benasahel, Nora 60
Bonaparte, Napoleon 47, 49, 91-92, 175
Bousquet, Antoine 2, 177-178
Brauer, Jurgen 125

Carr, Andrew 65, 109
Cavoli, Christopher (US Army, General) 84-86, 87
Center(s) of Gravity (COG) 40, 49, 121-123, 179
Challenge-Response Cycle 34, 55-56, 59, 61, 122, 186-187, 193
Clausewitz, Carl von 16, 20, 22, 26, 40, 49, 102, 119, 127, 191, 193, 197-198
Conflict realism 4-7, 12-13, 14, 27-28, 29-42, 125, 193, 195-196

Defeat mechanism 7, 46, 64-67
Dislocation 64, 66, 139, 174
Drone. *See* autonomous system
Drone warfare 2, 177-178, 181
Dupuy, Trevor 82, 126-128, 162, 165-166
Dominance 25-26, 61, 150-152, 156

Exhaustion 46, 53, 64-67, 79-81, 86, 90-91, 96, 103, 113, 123, 127-129, 132, 145-146

Fuller, J.F.C. 8-9, 20-21, 25, 41, 69, 83-84, 102, 109, 121, 125, 193
Gady, Franz-Stefan 19, 160

Hecht, Eado 64
Hoffman, Frank 64,
Hughes, Wayne 165

International Humanitarian Law (IHL) 127

Jomini, Antoine 49, 193

King, Anthony 46, 78, 114, 129, 157, 160
Kofman, Michael 19, 159-160, 170, 196

Leonhard, Robert 21, 61
Liddell Hart, Basil Henry (B.H.) 21, 41, 70
Lind, William 169
Luttwak, Edward 169

Mattis, James   183
Maneuver   111, 119
   Maneuver-Attrition debate   11
   Maneuver warfare   11, 18-19, 25, 62-63, 82, 85-87,
Meiser, Jeffrey   16
Mosul, battle of   188

Nagorno-Karabakh (War of 2020)   2, 4, 53, 83, 196
Nolan, Cathal   46, 86, 127, 145, 160, 163

Owen, William (Wilf)   60

Positional warfare   11, 19, 62, 82, 119
Power   6-7, 24, 29, 35-37, 39, 70, 79-81, 104, 121-122, 201
Principles of war   9-10, 21, 33, 69-96, 118
Principles of warfare   9, 33, 70, 71, 81-96, 118
Precision (as a concept)   46, 170
   Precision Guided Munitions (PGMs)   2, 10, 114-117, 152
   Precision strike   11-12, 14, 23, 30, 45, 52
   Precision strategy   42, 62, 84-85, 122
   Precision warfare   52-53, 175, 177-181
Precision Paradox   27, 32, 121, 174-177, 182, 185, 189, 191-193
Proxy war   2, 13, 14, 27, 33, 42, 45, 59, 63, 75, 105, 116-117, 130-137, 144, 149, 156, 183, 197

Rainey, James (US Army, General)   85
Rapid Dominance theory   23
Rogers, James   30, 177, 193
Roving warfare   11
Russia   10, 18, 31-32, 46, 48, 62, 74, 190
   First Chechen War   30-31, 87-90, 139-141
   Russo-Ukrainian War (Donbas Campaign)   73, 120, 130, 143, 149
   Russo-Ukrainian War (2022-)   4, 30, 52, 72-73, 79, 84-85, 99-102, 117, 125

Siege(s)   10, 13, 27, 42, 98-99, 103-104, 125-134, 147, 198, 200
   Characteristics of   147-151
   Logic of sieges   126-129
   Misconceptions   141-147
   Siege of Bakhmut (2023)   131
   Siege of Debal'tseve (2015)   120
   Siege of Donetsk Airport (2014-2015)   120, 130
   Siege of Fallujah (2004)   110
   Siege of Gaza (2023-)   130
   Siege of Homs (May 2011-May 2014)   133-134
   Siege of Luhansk Airport (2014)   120
   Siege of Mosul (October 2016-July 2017)   38, 117, 129
   Siege of Mariupol (2022)   99, 120, 130
   Siege of Vukovar (1991)   131-133

Systems theory 6-8, 65-66, 75-76, 166-168, 186. *See also* systems thinking
Svechin, Alexander 102, 170, 186

Taylor, Curtis (US Army, General) 60
Transparent battlefield 7-8, 46, 53, 55, 59-62, 65, 114, 201. *See also* battlefield transparency
Tuck, Christopher 71, 160, 163

Urban warfare 9, 98-123, 139-140, 155
   Principles of 118-123

Van Tuyll, Herbert 125
Victory 165

Watling, Jack 9, 19, 116-117
Watson, Joel 22

www.ingramcontent.com/pod-product-compliance
Lightning Source LLC
Chambersburg PA
CBHW041306110526
44590CB00028B/4259